RAISING ANIMALS FOR FUN & PROFIT

BY THE EDITORS OF COUNTRYSIDE MAGAZINE

TAB BOOKS Inc.
BLUE RIDGE SUMMIT, PA. 17214

FIRST EDITION

FIRST PRINTING

Copyright © 1984 by TAB BOOKS Inc.

Printed in the United States of America

Reproduction or publication of the content in any manner, without express permission of the publisher, is prohibited. No liability is assumed with respect to the use of the information herein.

Library of Congress Cataloging in Publication Data

Main entry under title:

Raising animals for fun and profit.

   Includes index.
   1. Domestic animals. I. Countryside.
SF65.2.R35   1984   636   83-24225
ISBN 0-8306-0666-1
ISBN 0-8306-1666-7 (pbk.)

Contents

Introduction   vi

1   Cattle   1
Sources for Cows—Longhorns—Shorthorns—Irish Dexter Cattle—Milking a Cow by Hand—Milking Tips—The Milking Machine and Its Use—Breeding the Cow—Energy Requirements of Cows Vary with Reproductive Cycle—Calving—Torsion of the Uterus—Doctoring Calves with a Stomach Tube—Caring for Your Sale Barn Calf—Treating a Sick Cow

2   Horses   36
Acquiring Draft Animals—Training a Cold—Harnessing a Horse—Hoof Care—Teeth—Energy and Feeds

3   Swine   61
Self-Feeder for Hogs—Hog Feed—A Model Feeding Program—Feeder Pigs—Summertime Hog Raising—Cold Weather Planning for Pigs—Plowing a Garden with Pigs—Artificial Insemination—Helping a Sow Give Birth—Selecting Replacement Gilts—The Science of Castrating Pigs—Hog Skinning—Butchering Pigs

4   Sheep   109
Blackface and Whiteface Breeds—Crossbreds—Equipment—Choosing the Right Breed—Finnsheep—Fencing Alternatives—Feeding Rack—Eating Habits—Worming—Hoof Trimming—Stopping Foot Rot—Finding the Right Ram—Length of Day Affects Breeding—Teasing Sheep—Managing Pregnant Ewes—Lambing—Post-Lambing Difficulties—Tender Loving Care and Milk Replacer Work Wonders—Docking and Castrating—Feeding Lambs—Maintaining Ewes—Shearing

| 5 | **Goats** | 168 |

Pygmy Goats—Angora Goats—Goat Barn and Hay Feeder Barns—Exercise Yard—Feeding Program—Drenching—Winter Milk Supply and Lighting—Breeding Basics—Kidding—Raising Kid Goats—Dehorning

| 6 | **Poultry** | 206 |

Anconas—Blue Andalusions—Brahmas—Cochins—Cornish—Dorkings—Jersey Giants—Langshans—Leghorns—Minorcas—Modern Games—New Hampshire Reds—Old English Games—Orpingtons—Plymouth Rocks—Rhode Island Reds—Sussex—Wyandottes—Geese—Muscovy Ducks—Pigeons—Turkeys—Quonset Hut—Chicken Feeder—Rations—Hatching Eggs—Bucket Incubator

| 7 | **Rabbits** | 252 |

Breeding Stock—Wood and Wire Hutches—Getting Ready for Summer—Winter Rabbit Raising—Breeding for Stamina

| 8 | **Bees** | 267 |

Package Bees—Examining and Handling Bees—Helping Your Colony Grow—Feeding the Colony—Increasing the Population of Colonies—Caring for Colonies in the Fall—Selective Breeding—Harvesting and Processing the Crop—Beeswax

Index   297

Also by the Authors from TAB Books Inc.

No. 1356 *Country Wisdom: The Art of Successful Homesteading*

Introduction

Homesteaders get a lot of pleasure and satisfaction out of raising animals. Considerable time and effort, though, are required to keep these animals healthy and productive.

This book has information on cattle, horses, swine, sheep, goats, poultry, rabbits, and bees. Topics include milking cows, training colts, skinning hogs, shearing sheep, dehorning goats, hatching eggs, building rabbit cages, and handling bees. Sections on housing, feeding, and breeding are particularly valuable.

All of the material in this book has been made available by the editors of *Countryside* magazine. Without their efforts and cooperation, this book would have been impossible.

Chapter 1

Cattle

Thinking of investing in a family cow? Whether buying from a neighbor or at an auction, understanding some of the more common dairy terms will help in choosing the proper cow for your situation as well as dealing with the person making the sale. Following is a brief glossary of frequently used terms to assist you in conversing with dairymen in a variety of situations.

Heifer. A young female cow from birth until of age to milk and beyond; sometimes up to the age of five years.

First Calf Heifer. A young cow who had one calf and is milking for the first time.

Dry Period. A time prior to the birth of another calf, usually six to eight weeks along, when the cow gives no milk and rests in preparation for another lactation.

Springer. A cow who is ready to give birth. When the cow's udder begins to swell, indicating she will calve within a week or 10 days, she is called a springer. Many dairymen feel this is the right time to buy, as the cow will adjust to her new home and the calf will be purchased with the cow. The price of a springer is often greater than that of a fresh cow. Other dairymen feel it is best to wait until a cow is fresh to buy, then any problem at calving time is past. The new owner only has to milk the cow.

Aged Cow. A cow who is past five years old.

When at an auction a ring man will often mouth a cow or look at her teeth to tell the age. These terms can be very confusing to the uninitiated. The basic ones are:

Baby Tooth Heifer. An animal with all baby teeth, usually under two years old.
Two Big Teeth. Two of the second teeth are in, indicating an age of two years.
Four Big Teeth. By this time the animal is three years or over and has four of her second teeth.
Six Big Teeth. The four-year-old heifer.
Coming Up on the Corners. An animal who is reaching maturity with the last of her second teeth coming in, usually at about age five years.
Full Mouth. This animal is probably five years or over.

Beyond full mouth the ring man will estimate age by the amount of wear on the teeth. Many ring men will never age a cow beyond seven years, though the animal may be over 10. The general appearance and condition of the udder should guide the individual making the purchase.

When purchasing a family cow, many things should be considered. The fawn-like little Jersey is a favorite choice. Jerseys are pretty, relatively small, give less milk but richer in cream, and have soft deer-like eyes that are very appealing. Jerseys can be the meanest cow in the barn.

After many years with all types of cattle, only the Jerseys have been really mean, especially the new mothers. Holsteins look gigantic next to a Jersey, but we prefer to sit under a larger, docile animal and milk rather than under a small, flighty one.

Guernseys and Brown Swiss are good milk producers. Though larger than the Jersey, they are quieter. Guernsey and Brown Swiss cows also produce richer milk that the Holstein does. Ayrshires tend to be more nervous than the other large breeds, but they can be good family cows.

SOURCES FOR COWS

There are several sources of family cows. A neighbor with a good herd is perhaps the best place to buy. If he is honest and keeps a clean healthy herd, the family would find that the purchased animal probably would live up to what the former owner predicted.

Auctions are the most common places for the purchase of cattle. A single farm auction is our favorite of the several types of auctions. In a commission auction, such as the weekly livestock markets, the chance of bringing home health problems is far greater as the cattle are run through the pens and alleys with cattle that are sometimes sick.

When a dairy farmer is selling out on the farm, the herd has usually been checked by a veterinarian and in some states must pass certain health regulations before being sold.

Many dairymen agree that the cow is the most temperamental, docile, stubborn, demanding, and yet beneficial animal that can be cared for on a farm. Dairymen work the grueling twice a day, seven day a week schedule because they appreciate and believe in the cow. She is an excellent teacher of responsibility, kindness, and perseverance.

LONGHORNS

Longhorn cattle, almost extinct in the 1900's have become more popular recently because they endure rough range conditions (Fig. 1-1). Many people predict that population growth will limit the cheap grain available for livestock feeding. When, that happens, cattle will be finished on ranges.

Longhorns are hardy, tough, and wily, but they are not difficult to handle.

"One thing I've noticed is that they forage a lot better than other cattle and that cold or hot weather don't seem to bother 'em as much," says Lauer Hieronymus, a commercial cattleman. "They'll eat stuff other cattle won't touch, and on poor range you can run maybe a third more Longhorns than you can other cattle. I guess it's

Fig. 1-1. A Longhorn is a tough animal.

because they'll eat weeds and things that other cattle won't.

"I'll tell you something else, too. They're not nearly as susceptible to disease as other cattle. You don't have nearly as much pink eye, and they don't seem to bloat and have the problems you have with more domesticated breeds."

Longhorn cattle breeders claim their cows are "calf-producing machines." They breed for a long time and quite often.

"On the first-crop heifers you get almost a 100 percent calf crop and you don't have to help 'em give birth either," he said. That's very important when you're running cattle on the range because beef producers can lose an estimated 6-10 percent of their potential profits from calving deaths.

Born small, maybe 45 to 50 pounds, Longhorn or Longhorn-cross calves get up and suckle quicker than any other cattle breed. The mothers are good milkers and the calves grow quickly, Hieronymus said.

"They'll raise a fat calf and they'll wean it at around six to seven months earlier than most other breeds I've seen", he said.

Cattle breeders also like Longhorns and Longhorn crossbreeds carcasses because they have little fat thickness. A study done by researchers at California Polytechnic Institute in the early 1970s shows Angus cattle finished in western states' feedlots had the best dressing percentage (62.9 percent). Longhorns were right behind with 61.1 percent.

Research on the Longhorn is limited. Universities like Texas A&M, Texas A&I, Oklahoma State University, and the University of Missouri are testing the Longhorn for its performance.

SHORTHORNS

Shorthorns are an old-fashioned breed. They perform equally well as both beef and dairy animals (Fig. 1-2). There are other so-called dual purpose breeds of cattle, but they only really excel in one field. Although there are really two types of Shorthorn cattle, Milking and Beef, there is a fine line dividing the two. Thus, it is quite easy to go from one to the other without changing over your herd. The Shorthorn was the first breed brought into this country. It was popular in Scotland and England for several centuries for both beef and dairy purposes.

Shorthorns are a gentle breed, being generally more tractable than the other beef breeds, as well as being tougher and hardier than the other dairy breeds. The mothering instinct has not been bred out, but the mother cow, when used to its owner, is usually not

Fig. 1-2. A Shorthorn cow is both an excellent beef and dairy animal.

overly protective of the calf (to the point of attacking the owner, as seen in some breeds).

While not giving the milk that a high-producing Holstein gives, they will not, on the other hand, eat (and manure) the amount that a Holstein will. A decent Milking Shorthorn cow can easily give from 20-40 pounds of milk having 3.6-4.8 percent butterfat test. A good to excellent cow can rival the dairy breeds for milk production records. If milk is needed, either for home use or for sale, the Shorthorn is right there to help out.

A Shorthorn cow will weigh from 900-1500 pounds and the bull from 1800-2500 pounds. The bull is usually tractable and very gentle, as compared to other dairy bulls. A well-fed yearling steer can weigh over 1000 pounds, often dressing 60 percent of nicely marbled choice or prime meat.

Shorthorns range in color from white through red roans to a dark liver red in color. When thinking of a pioneer's cow, the red roan Shorthorn immediately comes to mind.

They are excellent rustlers and are very winter hardy. When not kept in a warm barn, they grow a dense winter coat for protection against the cold.

The calves are easily delivered, as they are medium-sized at birth, putting on size after being born. There are some breeds that have 100-pound plus calves at birth. These are the breeds that have calving troubles, and the yearling sizes are no bigger than the Shorthorn when it comes to dressed weight.

The Shorthorn crosses well with any other breed of cattle. You can cross Guernsey cows with a good Shorthorn bull.

Even with the first cross, the Shorthorn blood shows up strong. The bull calves top the market as vealers or when fed out, being nice and blocky. The heifers show dairy characteristics, but yet are more beefy than their Guernsey dams.

Three acres of well-managed pasture and hayland can keep a Shorthorn cow for a year (figuring 1 acre for pasture, rotating half at a time and 2 acres for hay, planted in clover or alfalfa and grass mix) plus a little grain.

Right now there is a trend toward the new "exotic" breeds, especially in beef cattle. Everyone is hopping on the band wagon, breeding their cows to these bulls, by means of artificial insemination.

IRISH DEXTER CATTLE

Irish Dexter cattle are dual purpose, meaning they are good milkers and fatten well for slaughter (Fig. 1-3). They originated in western Ireland, where the soil is poor and the land was divided into little rental crofts (farms) of a few acres each. Pasture was poor and small—certainly no setting for a large cow giving many gallons on

Fig. 1-3. This Irish Dexter cow is three years old.

milk a day and in turn eating everything 4 acres could produce. About 200 years ago, someone referred to a small black cow, common in that area, which was kept outdoors on pasture all year.

"Dexter" is the Irish word for small and dark. There were terribly poor breeding practices in those days, and only in this century has any attempt been made to upgrade breeding practices and choice of breeding stock. As a result of inbreeding, etc., early statistics showed a rate of birth deformities in the Dexters that was higher than for other breeds. This reputation seems to be no longer desired, since the worst estimate would be that 10 percent of calves are born defective or dead. Only one person out of all the Irish Dexter owners we have met ever had that experience. This improvement is, no doubt, due to modern breeding practices now employed. At any rate, all breeds have their problems. Some have a high rate of sterility, others have serious complications in delivery, etc. These problems are largely avoided by buying a cow with a reputation as a proven breeder in any breed. The cost is greater.

A person with 3½-4 acres of pastureland could easily pasture two cows and their calves for about seven months with no supplementary feeding. It would be best if the pasture was divided down the middle with an electric fence. This allows the cattle to feed several weeks on one half while the other half grows. In the winter you can take the fence down and allow the cattle to roam over all the pasture picking over the summer rejects. If planned well, one cow will always be milking and the cost would be five-six months' supply of hay, plus a little corn.

MILKING A COW BY HAND

She may slap you alongside the ear with her manure-laden tail, she may heist a foot and put it down into an almost full pail of milk, or she might set that cloven foot on top of your bare foot. She could even kick you off the stool and spill all of the milk. These earth-shaking calamities don't have to happen if you train and outthink that dumb cow.

Cows come in many forms: some are hard milkers and some cows practically beg you to take the milk. Some come equipped with long teats and some with short. Take this physical conformations into consideration when buying a cow.

You will need a milk pail. It is much smaller at the bottom than at the top—and for a reason. It is made that way so that the pail will fit and rest comfortably between your legs. The bucket should not be set on the floor or ground while milking.

Fig. 1-4. You can make this anti-kicker hobble out of strapping and chain.

When you sit on the stool, your ankles will be closer together than your knees. The bottom of the pail can rest on your shoe or boot tops, above or below the angle depending on how long your legs are. Twenty pounds of milk can get heavy, squeezed between the knees.

The stool should be 12 inches high and wide enough to accommodate your rear end. If it is too high, your back will have to bend too much and it's tiring.

A one-legged stool can be used when there are many cows to milk, and that leaves both hands free. The one-legged stool has a strap between your legs fastened to another strap that goes around the waist, so that the stool just hangs from the buttocks, enabling you to set on it anytime you want without using a hand.

For a cow that does a lot of kicking or a young heifer not used to being milked or for letting a not-her-own calf nurse, an anti-kicker hobble comes in handy (Fig. 1-4). They are sold in feed or farm stores for a few dollars, but you can make one.

You'll need a piece of ⅝-inch chain, 18 inches long with a 1½-inch ring fastened to one end, and two pieces of lightweight metal strapping, 1¾ inches wide by 5 inches long. Cut a slot in one end and fasten one end of the chain to it. Bend the metal in the center until it is 1½ inches at the open end.

Take the other piece of metal and drill a ¾-inch hole 2 inches from one end and cut a slot ¾ inch long extending lengthways from the hole towards the end of the metal. The slot should be ¼ inch wide, allowing the hobble to be adjustable in length. Bend the metal as was done with the first piece.

The hobble is placed on the hind legs of the cow, just above the hocks (that part of the leg that would be a knee on a person; however, the hock bends in the opposite direction). The hobble half-circles each hind leg on the outside and the metal clasps hook over the large, narrow tendon above the hock. The chain is drawn fairly tight between the udder and the hind legs.

The cow's udder and teats must be washed and dried before each milking. You also wash your hands.

Sit up close to the cow with your head resting against her flank and your left knee propped against her right hind leg (and you do milk a cow from her right side).

Being close to the cow brings you in tune with her and allows you to feel the cow's next movement and take preventive action. If she kicks, there will not be room for her to really unwind.

The ideal position for the cow to stand while being milked is with the left hind foot about 6 inches farther forward than the right hind foot and the feet close together. That position puts some pressure on the udder and helps pour the milk down. Also, the right leg will not interfere with the movement of your left arm while milking the hindquarters.

If the cow does heist her leg or try to kick (if the hobble is not on her), a quick movement of your left arm between her legs and your hand grasping her left tendon above the hock will most often discourage her action.

Reading how to milk a cow is like reading how to ride a bicycle—you must do it yourself to get the feel and the rhythm. You can milk a cow by the crossways method, that is, by grasping her right hind teat with your left hand and her left front teat with your right hand. When those two are milked dry, switch to her left hind with your left hand and her right front with your right hand.

If the cow's teats are unusually short, a person will need to strip milk. This is done by grasping the teat with only the thumb and first finger and pulling straight down, letting the thumb and finger slide down her teat.

For normal milking, your hand reaches into her udder above the teat with a combination squeeze-pull down-squeeze movement. The hands are opening, reaching, and closing. The squeeze is started with the thumb and first finger and continued rapidly with the rest of the hand, all the while pulling down and slightly toward you.

Your wrists and arms play a leading role as you pull. Your arms drop with power, and that prevents the fingers and hands from tiring so quickly.

Milkers develop extremely strong fingers, hands, wrists, and forearms. Milking is a rhythmic motion—first one hand followed by the other (you will need to do it for the feel).

Nearly all cows have more milk in the hindquarters of their udder than they do in the front quarters. Perhaps a right-handed person has more grip in the right hand so that the frontquarters milk out first.

If a hindquarter has quite a lot of milk left in it, you can switch hands, with your right hand milking while the left hand squeezes the udder above the teat to bring the milk down faster.

In fly season the cow's switching tail can be irritating. Her tail can be held tightly under your knee with a tight squeeze of the leg, or the tail could even be tied to the cow's leg. (Some cows won't stand for that.)

You might be able to sweet-talk your partner into holding the cow's tail, or she or he may even milk the cow while you hold the tail. All the while those pesky flies are biting the cow like crazy.

After you get the hang of milking, it doesn't take much concentration and you can sit there day-dreaming. It is kind of peaceful.

MILKING TIPS

Milking is probably the most performed function on any farm with a cow or cows and is the job most often done by absolute amateurs. Whoever heard of a course on how to milk cows? Perhaps the cows would appreciate the results if a good sound course on milking were given to all those who have the chore of extracting milk from the cow.

The first consideration when milking the cow is that she is not a machine. She is a very feminine individual with feeling and emotions. Cold, rough hands can be devastating to her aplomb. A milking machine that is not working properly can leave the cow very wary of the milker.

To be a good milker, one must first understand the cow, her life-style, physical makeup, hormonal balance and emotional balance. The average cow is a creature of habit. Her day should progress along a normal routine of feeding, milking, exercise, and rest. Any major change in the routine to which the animal is accustomed may make a great deal of difference in milk production.

If a cow normally receives a dip of feed at 6 A.M., each day. The animal who normally is fed before milking will be excitable and hard to milk if the feed is withheld until the milking is finished. On the other hand, regular feeding an hour or so before milking or after the milking facilitates the milking as the cow finds eating more important than milking.

The milk cow is a docile creature and enjoys the company of

humans who respect her for what she is and do not attempt to train her with violence. The individual who continually beats the animal into submission is only creating greater problems. A kicking or excited cow will not relax for milking and the release of *oxytocin* from the pituitary gland. The oxytocin, when released into the blood stream of the animal, causes a letdown of the milk in the udder. This letdown of milk is best accomplished by kindly treatment and a ritual procedure of preparing for milking that stimulates this release of the hormone.

Even a change of milking personnel, on a day the farmer takes off, can upset the animal and lower production. A large crowd of visitors or a change of routine can lower the amount of milk produced.

A cow always appreciates gentleness in the handling of the udder. Immediately following the birth of a calf, the udder is especially tender to the touch. Even a milking machine that is not operating properly can irritate the udder. Any woman who has ever nursed a child knows how the animal feels. Perhaps that is why women down through the ages have been the "milkmaids." The word daughter is derived from a German word that originally meant "she who milks the cow."

The udder of the bovine is a very complex apparatus for the secretion of milk. Large supplies of blood are pumped to this area of the body and in passage through the udder, milk is produced. The milk in the udder is stored in small cells called *alveoli*. With the release of the oxytocin, the alveoli release their hold on the milk and allow it to drain toward into the teat canel. From the teat canal, the use of hands or the milking machine draws the milk out by opening the *sphincter* muscle, or rubber band-like muscle, that pulls the teat canal closed at the end of the teat.

Many cows, with proper stimulation, will start to "leak" milk. In these cases the sphincter muscles have relaxed sufficiently to let the milk come without the use of force.

The proper action for hand milking can be studied by allowing a newborn calf to suck on a finger. The calf milks the cow with a combination of sucking in the breath and caressing the teat with its tongue.

The actual hand milking is a combination muscular movement using the muscles of the forearm for the greatest amount of work. By gently grasping the teat in loosely bent fingers, squeeze with a gentle downward motion. While doing this, use the thumb in opposition and gently caress the teat also in a downward motion. Pulling, jerking, and squeezing with a hard pressure of the hand is simply not effective in bringing a full stream of milk.

The forearm should and will ache for a few days when a person first starts to milk a cow. With practice and a noticeable relaxation of muscles, the milking will become almost effortless if done properly.

THE MILKING MACHINE AND ITS USE

Machine milking is accomplished with the use of vacuum. The electric motor operates a pump that, when adjusted by a competent technician, will create the proper vacuum for operation of the milking machine.

This vacuum pump is established in a permanent location in the barn with pipe line around the barn above the stalls. A stall cock is located at each stall or between two stalls that, when open, allows the milker to be under vacuum.

The *pulsator*, heart of the milking machine, must be in good working order to milk the cow properly. As in hand milking, the pulsation of the milker creates a suction against the teat end as with a calf, interrupted by an almost caressing motion on the teat as the rubber inflation in the teat cup collapses and thus shuts off the suction action.

Pulsators come in many forms, depending on the make of the milking machine. All pulsators that work exclusively under vacuum have the same principle.

The air hoses to the teat cup shells are on one side of the pulsator and cause the collapse and opening of the inflation. The air hoses to the pail, on the other side of the pulsator, create the vacuum in the pail that sucks the milk from the teat when the inflations are open.

When the pulsator is placed under the pressure of the vacuum, the air is sucked out of the system, and the lid of the pail seals securely to the top of the pail. The pulsator then begins its pulsation.

The pulsator is usually a metal block with a series of holes connecting various parts of the machine. The vacuum sucks against the leathers until the leather unit moves towards the vacuum. As this is accomplished, the holes sucking the leathers are closed in the block. The holes on the opposite side are opened, causing the suction action to bring the leathers back to the first side. This pulsation, when directed into the proper air hoses and parts of the machine, causes the machine to pulsate and milk the cow.

As with any breathing apparatus, the minute openings of the pulsator must be kept very clean and dry (Fig. 1-5). The only area on most pulsators that use any oil are the compartments where the leathers work back and forth. The oil makes the pulsation easier and also creates a seal between the leathers and metal that will not let air by, thus insuring proper pulsation.

The more rapid the pulsation of a milking machine, the slower a cow will milk. A pulsation that is too slow can be harmful to the cow. An average 50-60 pulsations per minute is the optimum rate for good milking. This allows time enough between the pulsation for the milk to be let into the inflation and down into the machine, and yet does not suck long enough on the teat end to irritate the sphincter muscles and delicate tissue of the teat.

Proper care of the rubber inflations, the only part of a milker that has actual contact with the udder, is also a must for good milking. The inflations should be cleaned daily by washing and replaced before the rubber loses its elasticity and smoothness.

The vacuum line can also become clogged with milk, dirt, and anything else that an escaped vacuum can suck up into it. This line should be cleaned periodically by sucking a washing solution through it from the stall cock to the tank on the vacuum pump. Most vacuum tanks have an outlet to release this cleaning fluid.

With proper care and understanding of how a milking machine

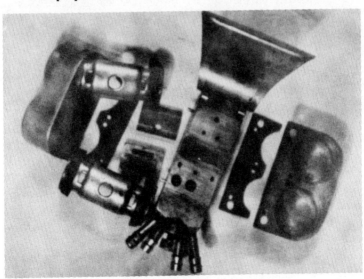

Fig. 1-5. As with any breathing apparatus, the minute opening of the pulsator must be kept very clean and dry.

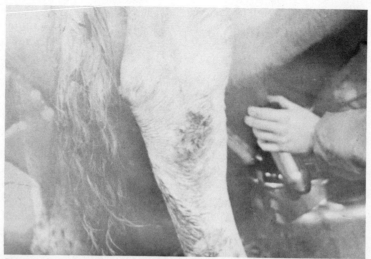

Fig. 1-6. A cow appreciates having the machine attached in an easy, gentle manner.

operates, a cow or cows can be milked for maximum production and good health.

When milking a cow, try to develop a rhythm of movement that will not startle the cow. Approach the cow quietly, speak softly to her, touch her on the hip, and make sure she is aware of your presence. Then and only then step in beside the cow.

A clanging milking machine can be upsetting to the unsuspecting cow. Handle the equipment so as not to annoy the animal. Attach the main air hose to the stall cock, turn on the vacuum, then set the machine down under the cow.

With a flighty cow, a cool head and slow deliberate movements are best. A raised hind leg can be very effectively stopped by the forearm, raised against the front side of the hock. Leaning a head against the flank of the cow will throw the cow off balance enough to stop many protests from hind feet.

A cow appreciates having the machine attached in an easy, gentle manner (Fig. 1-6). Place your hand at the top of the teat cup and break the suction as the inflation slides up onto the teat. Many milkers hold the teat cup at its base, and we have seen startled cows jump as the vacuum suddenly grabs hold of a teat.

Having the hand at the top of the teat cup insures against doubling the teat in the inflation. The cow cannot possibly milk this way, and it is very painful to the animal.

When removing a vacuum milker, always release the vacuum

before pulling it off the cow's udder. Stretching the teats out long and pulling the machine off forcibly with the vacuum still sucking full force not only hurts, but can damage the tender tissues of the cow's udder. Any damage that destroys the alveoli glands in the bag also destroys some of the cow's ability to produce milk.

Actually, the well-trained cow starts with the training of the first calf heifer. Improperly started in the production of milk, the heifer can be a menace rather than a docile producer of milk.

Once again, the first requirement is an understanding of the emotions of the heifer and judicial approach to her. If possible, be around the heifer before the calf is born. Always speak softly but firmly to the animal and try to handle her, petting her back, scratching behind the ears, and in general making her familiar with the hands of a human.

As the time of calving approaches, check the udder, always talking softly to the animal. Make sure all the teat canals are open. Each teat should produce a "waxy" try.

Having help available is best for the first few milkings. Our own method has the milker approach the animal, speaking to her and placing a calming hand on the hip. If the animal is curious about the milking machine, let her look it over and smell it. The helper always stands at the rear of the heifer, quiet and waiting.

The milker gently puts the machine in place and onto the udder. If the heifer protests with a foot, the milker first speaks firmly and may stop the leg with his forearm. If the protesting continues, it is time for the helper to place a hand on the pelvic area of the animal and gently scratch or pet the heifer.

Often this is all that is needed to get the animals through the first milking. For the heifer who decides that kicking is the only way to get rid of the milking machine, the helper then, without trying to hurt the animal, lifts the tail up and directly over the pelvic area.

The animal in this position has difficulty kicking. A stubborn heifer may need a little pressure on the tail by pushing it forward. Most animals will stop the kicking, however, and as the heifer settles down the tail is slowly released.

By always being very quiet, never beating, and usually only this small amount of persuasion, we can often put a milker on a heifer and let her stand alone while we go about other milking chores by the third or fourth milking.

To beat or yell at the frightened animal will only increase the fear and stubbornness. Always remember in any training of a new heifer, the animal is alive, breathing individual with a brain and a memory. The animal that is handled with a calm resolute manner will produce the greatest amount of milk over the years and will probably last much longer as a milk cow.

BREEDING THE COW

Cows reach puberty (age of reproduction) at nine to 13 months of age. The best age to breed is from 15 to 20 months, depending on the development of the animal to be bred. Cows come in heat on an average of 19 to 23 days and remain in heat anywhere from six to 10 hours, though the average is 16 to 20 hours.

Indications of heat are: the cow will (if pastured with other cows) attempt to ride other cows, will bawl loudly and frequently, will be restless and excitable, and will have a clear mucus discharge from a red swollen vulva. Not all cows will show this discharge. She will stand to be ridden by other cows, and her milk flow will drop noticeably. Some cows are very aggressive and bullish when in heat. For those novice cow handlers, never turn your back on a cow that's in heat. We've seen two cases where an aggressive, bullish cow attempted to ride an unsuspecting human.

The most preferred method of breeding cows these days is with *artificial insemination*. Natural service is confined, more or less, to the bigger beef operations where the cattle are raised under range conditions. There are several reasons why artificial insemination is preferred. Probably the basic reason is because it gives the farmer a large selection of bulls from which to choose the bull that will best complement each individual cow. In this manner, he can employ corrective breeding and improve his herd both in terms of production and comformation. Also, he isn't saddled with the expense and trouble of maintaining a herd sire. The average farmer with 45 to 50 head of cattle to breed does not want to bother with an unpredictable, potentially dangerous animal. With artificial insemination he can breed his cows to top bulls for an average price of $8 to $10 and is usually (though not always) entitled to two free repeats if the cow does not conceive on the first service. Also, with artificial insemination animals are not exposed to infection like they are when bred naturally. A bull has to breed only one infected cow, and he will pass the infection on through the entire herd.

Breeders associations maintain stud farms that are kept very clean. The bulls are tested for disease periodically and are routinely vaccinated. They cannot afford to put out inferior semen if they are to stay in business. Most stud farms allows visitors to see the bulls, and we would heartily recommend that this be done.

The only disadvantage to breeding with artificial insemination is that for the novice cow handler, it is sometimes very difficult to tell when the cow is in heat. Once heat is detected, then it's time to call the artificial insemination technician.

Technicians run on a very busy schedule. Calls received before noon are served the same day. Calls received after noon are served the following morning. Since cows have a very short heat cycle, it's important that the call goes in as soon as possible so the technician will be able to breed the cow before she goes out of heat.

If you notice the cow passing a bright red bloody discharge before the technician arrives, it's too late to breed—she has already ovulated. Note to novice cow handlers—if you breed your cow and see this discharge six to eight hours later, don't panic. It's normal.

Some things the artificial insemination technician will want to know when you call is: your name and address, the breed of cow, when you first noticed her in heat, and the breed and code number of the bull you want the cow bred to. Also, you should find out if you are entitled to a free repeat service. If the cow is registered, get a certificate from the technician showing all the information, as you'll need it to register the calf.

There are several breeding co-ops. Most have representative technicians throughout the country. Sire directories are available free of charge. When writing for them, state what breed or breeds of bull you are interested in. Most co-ops have separate directories for the different breeds. You might also ask for the name of the nearest artificial insemination technician. (A local farmer could also tell you, as could the vet or feed dealer.) The price list you'll receive from an artificial insemination co-op will be for semen only, so add the technician's service fee to find out what the total cost will be.

For sire directories write to: American Breeders Service, De Forest, WI 53532; Select Sires, Inc., R. 3 Box 126, Plain City, OH 43064; Atlantic Breeders Cooperative, 1575 Apollo Drive, Lancaster, PA 17604; Sire Power, Inc., R. 2, Tunkhannock, PA 18657; Curtiss Breeding Service, Cary, IL 60013.

ENERGY REQUIREMENTS OF COWS VARY WITH REPRODUCTIVE CYCLE

The energy requirements of the brood cow vary tremendously at different stages of the reproductive cycle, according to Gerry Kuhl, extension livestock specialist at South Dakota State University.

Kuhl breaks the reproductive cycle into four periods, for the benefit of showing nutrient needs: mid-gestation, 60 to 90 days before calving, calving through rebreeding, and the end of breeding to weaning.

Mid-Gestation

In mid-gestation, which typically occurs during the early wintering period, nutrient requirements (including energy) of the brood cow are at their lowest point. The major biological function is simply maintenance.

This is the time, says Kuhl, to take advantage of poor quality feeds, such as grazing winter range and feeding poor to medium-quality hays, straws, and other harvested crop residues. These feedstuffs can supply a major portion of the cow's energy requirement when properly supplemented.

Before Calving

In the last 60 to 90 days before calving, the cow is in a critical period of the reproductive cycle, second only to the calving period in importance. In this time adequate energy is needed for rapid fetal growth, in addition to maintenance. About 80 percent of the total fetal development will occur in the last two months of gestation, and the cow is preparing for lactation. Consequently, cows should normally be gaining 0.8 to 1 pound per day, depending on body condition.

Through Calving

The period of calving through rebreeding is the most critical stage in the reproductive cycle, says Kuhl, with energy requirements at their peak. The average cow needs about 50 percent more feed intake, 70 percent more energy, and more than 100 percent more protein during this period than when dry. The cow loses about 120 to 140 pounds at calving. This weight should be regained within 90 to 100 days after calving. In addition, she has to produce adequate milk for the calf and get her reproductive tract back in shape for rebreeding and also meet her maintenance requirements.

While cows are generally on new spring pasture during this season, an additional energy source may be profitable. If the cows are not in good gaining condition at this time, supplementation with grain or other high-quality feeds will often pay dividends in terms of increased reproductive performance.

After Breeding

From the end of breeding to weaning, energy for milk production and maintenance are still required, but the critical feeding period is past once the cow is rebred. Most beef cows will taper off in milk production, and the calves will be consuming other feeds in addition to milk.

The cow's energy demands are lower at this stage than during early lactation. Therefore, says Kuhl, inadequate nutrition at this point will primarily affect only milk production and consequently her calf's rate of gain. If the feed supply is short at this time, creep feeding the calf is the practical way to boost weaning weights. Underfeeding at this time will generally not have any detrimental effect on her fetus, however, since its growth rate is very slow during the early stage of preganancy.

CALVING

About eight months after being bred, the cow will begin to build her udder. (Remember, if she has had a calf last year and was milking, you dried her up eight weeks before she was due to calve.) Some cows will do this sooner, and others wait until the day they calve, so don't worry if she is only two days away from calving and hasn't built up much of an udder.

From the time one week before she is due to calve, and one week following calving, she should be kept in a roomy box stall in the winter or a small, grassy lot during the summer. There are several reasons for this. In the first place, it is natural for the cow to try to go off in some quiet, secluded spot to have her calf. When her time comes, she will just quietly fade away (even over a fence) to this place. Should she have trouble, you wouldn't know. Many times serious trouble could have been easily averted by just keeping an eye on the cow at calving time.

Some cows will come down with milk fever just prior to calving or just after. This condition is found in high-producing cows (very rarely heifers) of milking breeds, brought about by a low blood calcium. It shows up with an ascending paralysis, that is, paralysis beginning with the hind legs and working forward, until the cow goes into a coma and eventually dies. It is usually easily treated by giving a calcium-phosphorous solution intravenously. It must be discovered in order for your veterinarian to arrive in time. Keep milk fever in mind any time a good cow cannot or will not get up or is staggery at calving time. It is not possible to even suspect milk fever if the cow is somewhere out there in the woods.

Not all "down" cows after calving have milk fever. Once in a while, a cow will have a nerve pinched during calving and be unable or unwilling to get up. This is the prime reason for having a roomy box stall for her to calve in. A cow that calves in a stanchion, especially on cement, and pinches a nerve will be very awkward. (It acts and feels as though her hind legs were asleep.) Many cows such as this end up falling and fracturing a leg or pelvis.

Just before calving, you will notice a marked change in her udder. It may have been large a week ago, but now it will spring out, tight and extra full. As the actual time of calving is reached, she will have a very relaxed vulva. The first signs of actual calving you will see are the long pinkish strings of mucus. Don't confuse these with the normal clear mucus a cow often passes after standing up. Many times she will appear cramped and shift weight from one foot to the other. Her tail will arch as actual labor begins.

Normal Calving

There are two positions that a calf can be born in that are considered normal. The most common position seen at birth is the *front presentation*, with the feet of the front legs closely followed by the head (Fig. 1-7). It appears that the calf is diving out into the world. Should the calf get stuck while in this position, it is usually

Fig. 1-7. A calf is coming in unaided in front presentation (photo by Jackie Spaulding).

only necessary to provide a little extra downward pull (slow and steady) to aid the cow in expelling the calf. Be sure the head and both the legs are in position. If they aren't, you won't be able to aid the cow and may really foul things up badly.

Another position often seen, and considered normal, is the *rear presentation*. Here the hind feet are presented first. You will be able to tell this presentation from the position of the feet, upside down, with the bottoms of the feet showing. Following the feet will be the hocks and tail. The one danger here is that when the umbilical cord breaks, the calf will begin to breathe. If his head is still in the birth canal, he may inhale fluids or smother. If you are present, give a little added pull when the calf is halfway out, so that he gets the first breath in the open air. Be careful here to catch the head. It is a long drop to the floor—especially concrete.

Should the calf inhale some fluid or not be breathing well, it will be necessary to clean out the mouth and nose. Sticking a finger up his nose will cause the calf to gasp and sneeze, which will start breathing and clear the breathing passages. Should this fail, smack the calf smartly on the side of the chest. The shock and the compression will help start breathing. Remember, a human doctor smacks a baby on the rump a good crack at birth. You won't hurt the calf, who weighs 10 to 20 times more than a baby.

Abnormal Calving

Any cow or heifer that has been in labor for two hours without observable progress should be closely examined. Cleaniless is essential in the examination to prevent infections. If the calf position is normal and the size not too large for the cow, wait another hour or two before giving assistance.

A difficult birth (*dystocia*) can occur for several reasons: abnormal fetal position, excessive calf size, small pelvis size of the dam, abnormal (damaged) pelvis of the dam, abnormal fetus, undilated cervix, lack of uterine contraction, twinning, or a twisted uterus.

The most common cause of dystocia is a misplaced leg or head. A calf cannot be born unless each part of the body is in the right place. If one leg is back or the head has been shoved down below the brim of the pelvis, it cannot be born.

Never put undue pull on any calf, especially with a tractor or fence stretcher, and especially if you haven't checked the position of the fetus. Even when the calf is in position, it is a dangerous idea to use these methods. Torn and ruptured uteruses, broken pelvises, and broken legs on calves are often a result.

If a cow has been working for over an hour and has made no progress, soap up well and go in for an examination. You should feel two legs, both belonging to the same calf, and either the head or the tail, depending on the presentation. If one leg or the head is back, you can try to bring it into the right position. This can often be done by pushing back on the calf, and at the same time, bringing the leg or head around. (The leg is usually much easier to get than the head.) If you work 15 minutes and make no progress, call your veterinarian. If you work too long, you will tire the cow out, chance losing the calf, and maybe even throw the cow into shock.

In a case where the calf is too large for the cow, you should call your veterinarian immediately. Usually it is possible to pull it by means of a special calf jack. This not only pulls on the calf, but pushes against the cow's buttocks, and allows just the right angle of pull. It is a slow, steady pull, much more successful than having four men or a tractor pull.

In rare cases, it may be necessary to do a *caesarean*. Here it is usually the farmer's own fault, as it is usually a case of a five or six-month-old heifer having been bred or a small breed cow, such as Jersey, having been bred to a large-calf breed, such as Simmental.

Many complicating factors can come up, such as a dead fetus, which absorbs fluids and swells up much too large to be expelled, or monstrosities, such as a calf with two heads, stiff legs, water head, etc.

The average cow probably will experience none of these difficulties and will go ahead and calve normally. Just be sure to have her where she can be watched and have her on good footing in a clean area.

Be sure to bathe the calf's naval in iodine soon after birth. Such troubles as scours, swollen joints, and arthritis can often be prevented this way.

TORSION OF THE UTERUS

Torsion of the uterus is not uncommon, but usually the degree of twist is not great, and many mild torsions correct themselves during the birth process. Torsion of the uterus is most common in dairy cattle that are confined, but it also happens occasionally to beef cattle at pasture.

There are many possible causes for uterine torsion. In advanced pregnancy, the largest part of the uterus lies free in the abdominal floor and supported by the rumen, internal organs, and abdominal walls. If the nonpregnant horn of the uterus is small or nonfunctional, the instability of the uterus is increased. This fact, plus the way the cow gets up and down (rising on her hind legs first when she gets up, and lowering herself on her front legs first when she lies down), suspends the uterus in the abdominal cavity, almost swinging freely. Any sudden slip or fall can create a torsion.

Other factors that can increase chances of torsion are lack of muscle tone in the uterus, lack of fluids around the fetus in the uterus, sudden violence such as falling, rolling or fighting, and strong movements of the fetus.

Uterine torsion is seen most often in advanced pregnancy, but it can occur earlier. Torsions of 180 degrees (half circle) or less may be present for days or weeks without symptoms, until the cow goes into labor and cannot deliver her calf. Torsions of 45 degrees or 90 degrees are quite common during pregnancy but usually correct themselves just before or during birth. A torsion of 90 degrees or more will usually cause problems when the cow goes into labor.

In rare cases, the torsion may be 180 degrees to 360 degrees, twisting the birth canal tightly closed. These severe torsions also restrict the blood supply to the fetus and usually cause the death of the fetus unless the torsion is corrected soon. In neglected cases, the uterus, vagina, or one of the large uterine blood vessels may rupture, causing severe bleeding into the abdominal cavity.

Symptoms of uterine torsion before the cow goes into labor may be completely lacking if the torsion is mild (45 degrees to 90 degrees or even up to 180 degrees). When the torsion is 180 degrees or more, the cow will show definite signs of abdominal pain. She usually loses her appetite, becomes constipated, stops chewing her cud, had a rapid pulse rate, and becomes restless or colicky—kicking at belly or switching her tail. If a cover over six months pregnant shows these symptoms, she should be checked by a veterinarian.

In severe torsions that are not diagnosed and treated early, the fetus usually dies. It can also kill the cow. Severe torsions that occur during labor (usually during the last part of early labor or the first part of active labor, when the fetus is often quite active) will have a better outcome if the cow is helped before the blood supply to the fetus is completely restricted. If the cow is assisted soon enough, a live calf can still be delivered.

The problem with detecting uterine torsion that occurs at birth is that the symptoms are often so mild that it just looks as though the cow is merely in the first stages of labor. The cow is uneasy, restless, and may kick at her belly and switch her tail. She doesn't get down to the business of serious labor. She doesn't start actively straining because the twisted birth canal prevents the fetus from starting through the pelvis. Entrance of the fetus into the pelvis is what stimulates the cow to start abdominal straining. A prolonged early labor without actual straining is a characteristic sign of uterine torsion.

The fetus is usually sideways or upside down because of the rotation of the uterus and may be dead if the cow has been in labor very long. In all cases of sideways or upside down presentations, the cow should be checked for uterine torsion and signs of rupturing. Determine the direction of the torsion before trying to correct it.

Chances of recovery in a cow with uterine torsion are fair to good if the torsion is discovered and corrected early, before the uterus ruptures and peritonitis sets in. The chances of delivering a live calf are somewhat lower unless the torsion is discovered quickly and corrected before the fetus dies.

Correction of a torsion can be accomplished in several ways—by a caesarean operation, through an incision in the cow's flank by rolling the cow, or by rotation of the fetus and uterus through the birth canal. Rolling the cow is the oldest and simplest method, but requires the help of several strong men. The object in rolling the cow is to rotate her body in the same direction as the torsion, rolling her rapidly enough to overtake the more slowly rotating fetus and uterus.

When rolling the cow, she should preferably be out of doors on a gentle slope with her head lower than her rear quarters. After she is down, her two hind legs should be tied together and her two front legs tied together, leaving 8 to 10 feet of rope on each tie for pulling. Do not tie the front and hind feet to each other, for this tends to compress the abdominal cavity, making the uterus rotate with the cow. The cow's head should be held extended by a halter and rope. The cow should be laid on her side on the same side as the direction of the torsion. She should be rapidly rolled over by a strong pull on the ropes fastened to her front and rear legs. After she has been rolled over 180 degrees, her body must then be either rolled back very slowly to the original position, or pushed over her legs and belly so that she is once more lying on her side, ready to be rapidly turned over again.

After two or three rapid rotations of the cow, the birth canal should be examined again to see if the torsion is corrected. If it is corrected, the spiral folds and constriction should have disappeared and the calf should be readily felt with the hand. If the cow is rolled in the wrong direction, the torsion will be made worse.

If the torsion occurs at the time of birth and is less than 240 degrees, it can sometimes be corrected by reaching into the twisted birth canal, getting ahold of the fetus, and rotating the fetus. In order to accomplish this, the cow must be standing up, and it helps if she is not actively straining. Sometimes a shot of epidural anesthetic must be given to keep her from straining. If the membranes around the calf have not yet broken, these should be ruptured and the fluids let out to reduce the weight and size of the uterus. The leg of the calf should be grasped below the knee, and the knee flexed. By pushing back on the fetus and twisting the leg in a rotating manner at regular intervals, the fetus and uterus are rocked back and forth in an arc of about 10 to 12 inches. Then with a sudden strong twisting and lifting motion, the torsion can sometimes be corrected. In difficult torsion cases, a caesarean section may be necessary.

After the calf is removed, the uterus and birth canal should be carefully examined for rupture, retained placenta, and internal bleeding. The cow should be given antibiotics if any trauma to the uterus or birth canal has occurred.

DOCTORING CALVES WITH A STOMACH TUBE

A stomach tube—a simple plastic tube about 5 feet long—goes into the calf's nostril, down his throat, and into his stomach. It is a quick and easy way to get medications and extra fluids into a dehydrated or debilitated calf.

We've also used it to save a 400-pound calf with coccidiosis (we had to feed him three times a day for 21 days with the tube while he was too weak to eat, but he was tough and pulled through), several calves that ate dirt and got "plugged up" (used the tube to wash out their stomachs—running in clear water, then letting the dirty water back out), and other cases that needed light nourishing food during recovery from various problems such as quick pneumonia, *infectious bovine rhinotrachaeitis* (IBR), and one calf removing from abdominal surgery.

We've also used the tube to get gas out of several bloated calves. The air rushes out the tube if you can get it placed correctly in the stomach.

Another handy use of the tube is to give colostrum to newborn

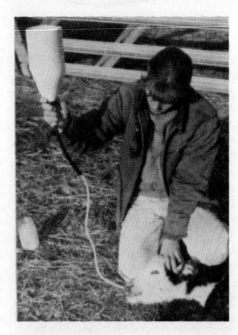

Fig. 1-8. Giving colostrum to a sickly newborn calf.

calves that are weak, cold, or for some other reason don't nurse right away (Fig. 1-8). We freeze several gallons of colostrum a year and store it in the freezer in pint-size plastic containers that can be easily thawed by setting in warm water. (Don't thaw colostrum rapidly; if you heat it too fast, it curdles and loses much of its value).

Our biggest use of the tube is in treating scours. Dehydration is one of the primary problems when a young calf scours, and the tube is the easiest way to replace those fluids. You can do all you can to eliminate the cause of diarrhea (using the proper antibiotics, etc.), but if he becomes too dehydrated he may still die. Not only does he need fluid to replace what he is losing, but he also needs plenty of liquid to insure best results from those medications you are giving him. Some antibiotics (especially sulfas) can be hard on the calf (injuring kidneys) if there is not enough fluid in his system to properly handle them and eliminate them from his body.

When we get a case of scours, we take the calf away from the cow (we have a small barn with individual calf stalls for sick cases) for 24 hours and feed the calf medicated liquid and electrolytes every six hours, then milk out the cow and put the pair together again. At first glance this looks like a lot of work, doctoring the calf every six hours and milking out the cow, but it is actually time-saving. Calves doctored this way usually recover within 24 to 36 hours, whereas the calves we doctored the conventional way (giving them pills once or twice a day and leaving them with their mothers) often dragged on sick for a week or more. Taking the calf away from the cow gives his irritated digestive tract time to begin recovery.

Revolutionized Calf Doctoring

The stomach tube revolutionized our calf doctoring. Before we learned about it we fed sick calves extra fluid with a nursing bottle, but it took a lot of time and patience, and with a stubborn calf that didn't want to drink there was always the risk of choking him. The stomach tube solved that problem. It is the quickest, easiest, and safest way to get liquid into a sick calf. The liquid we use is just plain warm water, with a teaspoonful of powdered electrolytes in it. (You can get this powder from your vet.) These are the salts and minerals the calf loses when he scours; by replacing these you tend to keep his body chemistry in proper balance.

When we treat a scouring calf, we give him electrolytes in about a quart of warm water every six hours. We add other things to this water, too, depending on the calf and what we are treating him for—antibiotics, preparations to soothe and coat the irritated digestive tract and to slow down intestinal contents, protein, or other forms of concentrated nutrition for a weak calf, etc. Even castor oil or mineral oil will run through the tube if it is warm (we shake it up with very warm water).

The tube is a flexible piece of plastic about 5 feet long. Its diameter should be large enough so that liquid will run through easily, but not so large that it irritates the calf's nose and throat. A good size is 5/16-inch external diameter 3/16-inch internal diameter). Smooth one end of the tube so it won't be sharp-edged. You can buff and smooth it on a grinder, whittle it with a knife, or use heat.

A good way to attach a funnel to the other end is to put an ordinary kitchen funnel into a small section of flexible gas line (5/16 internal diameter) or fasten the mouth of a plastic jug with the bottom cut out to the gas line. The other end of the gas line can be fitted over your tube when you are ready to pour the liquid.

The liquid should be body temperature, which is 101 degrees Fahrenheit. We like to have the tube in a jar of hot water until we are ready to use it. When it is warm, it is flexible and easier on the calf's nose. Shake out all the water. A tiny bit of Vaseline petroleum jelly on the tip of the tube makes it go down easier without hurting the calf's passage.

Inserting the Tube

Putting the tube into the stomach is not difficult. A calf can be doctored this way in less than five minutes. The calf can be either standing or lying. When you get the tube in the nostril and started down, you will have to feel when it gets to the back of the mouth (about 6 inches of tubing) and let the calf swallow it. If you don't give him a chance to swallow it, the tube will go down his windpipe and into his lungs instead of down his esophagus and into his stomach. Some calves swallow the tube readily, and others you must take more time with. You can usually tell when the calf swallows it, and as you feel him swallowing ease the tube on down. You have to provide the momentum as he swallows to move the tube along.

As you push the tube on down, you will get indications as to whether or not it is in the proper passage. If the calf coughs or gags, it is in his windpipe and you'll have to take it out and start over. If the tube goes down into his stomach, you can usually smell a foul odor if you sniff your end of the tube. Sometimes some of the stomach fluid will start back up the tube and then you'll know for sure it is in the stomach. When the tube goes into the stomach, it usually goes down easily and doesn't stop.

If the tube goes only part way down, you probably have it in the windpipe. If you are still in doubt as to whether it is in the proper place after the tube is down, blow on the tube. If you hear gurgling noises, it is in the stomach. If it makes the calf cough, it is in his windpipe.

After the tube is in the proper place, attach the piece that has the funnel and pour the liquid down, keeping a steady flow in the tube. When you are through, you can detach the funnel piece and blow the last bit down. Then put your thumb over the end of the tube and pull it quickly but gently out, so no fluid will leak on the way.

If it is a scours case we're treating, we milk out the cow after we've dosed the calf four or five times and put them together. If we have a stubborn case, we give the calf additional fluids for 12 to 24 more hours after he's back with the cow.

On our ranch we try hard to save every calf. Discovering the sick ones early is very important, and the stomach tube has proved to be one of the best tools with which to "doctor" them.

CARING FOR YOUR SALE BARN CALF

It's a lot easier to find the type of calf that you are looking for at a livestock sale barn, where all types of livestock are auctioned off, than from a private individual, which is the best source of healthy calves that stay that way. If you have no particular type of calf in mind, it is usually possible to go to a nearby farmer and either reserve a calf, yet unborn, or find one already in a pen, seemingly just waiting to go home with you. These calves may cost a bit more than one at the sale barn, but there are better chances that they will be healthier.

If you have a preference for offbeat breeds, such as Milking Shorthorn, Jersey or Guernsey, instead of the more usual Hereford, Angus, or Holstein, you may have a harder time locating a calf, especially a heifer. The chances of finding just the right calf are much greater at the sale barn, as perhaps 400 calves pass through each sale. Maybe there are 399 Holsteins and one Jersey, but one is better than none. Even if you are looking for a Holstein bull calf for a beef steer, the thrill of buying a bargain at auction may appeal to you. You can find real bargain calves at a sale barn, but the trick is being able to tell the bargain from the lemon.

Bargain or Lemon?

Be aware that there are two groups of calves at most sales: *deacons* (newly born calves) and *veaelers* (calves that are larger, up to about 250 pounds, sold by the pound for veal). Veal calves are always quite expensive. The deacons, usually sold by the head, are more moderately priced.

The best thing to do is to go to at least one sale, with your checkbook and wallet held firmly by another uninterested party. Make up your mind that that trip is just for educational purposes. Learn: how calves are sold, to understand the auctioneer, to understand how he works (does he start high, and if he receives no bid, does he lower the bid, or the opposite?), his slang for calfhood illnesses or health problems (slow may mean sick, a pimple may mean a rupture, wet tail may mean scours, spot in the eye blind, etc.), why meat buyers bid less for certain calves, and how to spot sick calves. It is usually best to go with someone very familiar with calves and their health problems to help you learn.

The first group of calves may bring quite a high price, especially if there are two or three buyers that are buying large numbers of calves. They are either meat or contract buyers who aren't out for "bargain" calves but, instead, want to fill their order and get on the road. It's usually quite easy to spot these buyers, as they are known and watched by the auctioneer, and they buy many calves—too many to be replacement heifers or to be raised for veal by one party.

Beware of any bargain calf that pops up while these men are still buying. They may let one go by that is a light breed, or very small for the breed, but often there is something wrong with the calf that makes it a risk, not a bargain.

After the big buyers have either slowed down on their bidding or gone home with their loads, you have left a very few bidders. Among these are few dogged meat or contract buyers, private individuals with lots of patience, and the scalpers—a group of buyers that only buy bargain animals, often after the majority of the other buyers have tired of the sale and have gone home.

Suddenly, the price of the average calf will drop. Unfortunately, though, for heifer buyers, the heifers are usually grouped separately and sold before the majority of buyers have left. At the end of the heifers, there can be a real buy, or there may be a small heifer or one of the light breed that Holstein buyers do not want.

When you are decided that now is the time to buy, pricewise, really keep an eye on the calves. Very cheap calves will often have health problems. Look for:

☐ Scours (severe diarrhea, often evidenced by a pasty tail or dirty hindquarters).

☐ Joint ill (infected knee joints evidenced by swollen knees)

☐ Umbilical infection (swollen umbilical area).

☐ Umbilical hernia (swollen umbilical area).

☐ Pneumonia (usually older calves—stunted growth, rapid breathing, head held outward).

☐ Dwarf Hereford calves (usually have bigger heads).

☐ Very newborn calves (have damp, red, umbilical cord, clumsy).

Let's say that you've found the calf. It comes into the ring, looks about, bounces around a little, has a clean rear end, smooth shining coat, no umbilical swellings, and looks to be about a week old. You found yourself bidding, holding your breath, bidding again (while putting a hex on the other bidder), then realizing that he has quit as the auctioneer says "Sold!" and points to you. Suddenly the little calf that is being chased from the sales ring is maturing to the greatest milk cow or the most tender meat that you have ever seen. Like a proud parent, you pay for the calf, then go to take it home.

Pluses and Minuses

By the time the calf has reached the sale barn and been placed in with other calves, he has received some strong minuses. These are stresses. First, the owner may have been planning to take the calf to the sale barn since it was born and did not bother to give the calf much (if any) of the vital colostrum milk (the first milk of the cow that gives the calf antibodies to fight disease). The calf may have been hauled to the sale barn in an open trailer in below freezing weather, after having spent a week tied in a very warm dairy barn, or having been shoved into the trunk of a car for a 50-mile trip or being shoved in with a bunch of 1000-pound cows to save hauling expenses.

Then the calf is penned in with several other calves, usually of the same size and sex. Among these calves are usually ones that have severe scours or other contagious conditions.

You have to add enough pluses to outweigh those heavy minuses. Be prepared to haul a calf home when you come to buy one. If it is hot outdoors, be sure the calf will have fresh air and not overheat. If it is cool or cold, keep the calf warm. This can be done by carrying the calf in an enclosed trailer bedded deeply with straw, an enclosed topper on a pickup, a special plywood calf crate, or even in the back seat or cab of the truck. If you do choose to carry the calf in the cab or in the back seat, have a helper ride with you, both to help restrain the calf (you don't want it jumping on your lap while driving down the interstate) and to help clean up calf messes.

Try very hard to locate some goat milk before buying the calf. Have it on hand, along with a calf bottle and a warm, small stall, when you arrive home.

After arrival quickly transfer the calf to the stall and give it a small amount of warmed milk. Do not feed it too much. A small feeding is a plus. A large feeding is a big minus, as the calf is upset from traveling, and a large feeding will upset its digestive system, resulting in scours. Plan on feeding the calf about 1 quart of warm milk every six hours for the first week, if at all possible.

Give numerous small feedings, with time enough in between for the milk to digest and the calf to remain hungry. A full calf is soon a sick calf.

After a week, you may feed 2 quarts every six hours. That amount should be increased slowly, not all at once.

Be always alert for scours—the number one killer of baby calves. Your chances of averting them is much greater if you use goat milk or cow's milk instead of powdered milk replacers. We have raised many calves and always had rotten luck with milk replacers until the calf is four weeks old or older, when it is out of the scour age and its digestive system is toughened up a bit.

Bed the calf's pen a little scantily. The normal calf will have several droppings a day, usually of a yellowish pasty consistency, but formed. If suddenly there are no droppings, be very suspicious. Check the bedding very carefully for any foul-smelling wet spots. Often scours start very suddenly, with the manure changes to the consistency of gravy, which oozes down through the bedding. If you can not spot any manure, make up your mind to watch the calf closely until it passes a stool.

If the calf has droppings that are a little loose, but not watery, just cut down on the milk given for two or three feedings, substituting warm water for the balance of the milk. This usually clears up nonbacterial scours. If not, drop all milk feedings and get some electrolyte formula, such as Lytren by Johnson and Johnson (available at most drugstores) and give it in place of the milk for one day. This often stops the irritation of the digestive tract, easing the scours, where the milk would continue the irritation and worsen the scours.

Should the calf have scours for longer than one day, and simply removing the milk does not bring about a recovery, get some kaolin-pectin with added neomycin, if possible, from your veterinarian. Give up to 6 ounces every two hours, depending on the size of the calf and the severity of the scours. Remember that you will be spilling some, so dress accordingly. Be careful in giving the medicine, so that you do not choke the calf. Usually if you place two fingers crosswise in the mouth, then pour the medicine over your hand, the calf will swallow readily, while mouthing your fingers, avoiding choking and fighting.

At the same time, give an electrolyte solution, either intraperitoneally or subcutaneously or orally. Your veterinarian can supply you with inexpensive disposable needles and tubes (which can be cleaned and be reused), along with the electrolytes. Electrolyte therapy is very important, as not only does the scouring calf lose body fluids, but also necessary body salts and chemicals, which the electrolyte solution replaces in the correct balance. Without these, the calf will soon die.

It is very important to keep the scouring calf warm, even if added heat in the form of a heat lamp, back porch (heated), or heating pad (protected) is needed. A scouring calf may weaken and become unable to produce enough body heat to maintain a normal body temperature. If the calf has a subnormal temperature very long, it will weaken severely and die, not matter what drugs are given.

With kaolin-pectin and neomycin, given every two hours, and electrolytes, along with a good, warm stall and watchful care, the calf can usually be pulled out of the scours within a day's time. If not, contact your veterinarian immediately, telling him exactly what you have done for the calf. Sometimes additional antibiotics or sulfas must be given, along with intravenous electrolytes.

After the Pluses Outweigh the Minuses

As the calf grows, he will need a supplement in addition to milk. We have had the best luck with Carnation Alber's Calf Manna, fed per directions. We do not like medicated calf pellets or growers. At a few days of age, the calf will chew the Calf Manna, and by two weeks, it will be eating a cupful.

As the calf eats more Calf Manna, you can add a good mixed calf growing mix (they usually have cracked corn, rolled oats, soybean oil meal, molasses, etc.), fed free choice. If you have plenty of milk, the calf can be fed increasing amounts of it, along with the grain, and the hay it begins to eat. If possible, feed first hay that has been in front of older cattle or goats, as they innoculate the hay with rumen bacteria from their mouths as they eat. Calves fed hay fresh from the bale have a tendency to get pot bellied, as their rumen doesn't develop to utilize the hay very well.

If you do not have unlimited milk, you can either wean the calf at four weeks, providing it is eating both grain and Calf Manna very well and is adapting to hay, or switch it carefully to a good powdered milk replacer, containing milk products and not soy flour, wheat flour, etc.

From four weeks on, just watch the calf for signs of pneumonia (loss of appetite, rapid, shallow breathing, and depression along with an elevated temperature). Soon it will be that super cow you dreamed of as it left the sale ring.

TREATING A SICK COW

How does one know when a cow is really sick and a call to the veterinarian is in order? Take milk fever for an example. This actually is not a fever but a depression of temperature and usually occurs within a few hours after the cow has calved.

The major cause of milk fever is an improper ratio of calcium to phosphorus in the body of the animal. The growing calf depletes the body calcium of the cow. With the onset of milk production, the cow can become sick or even die in a very short time.

The cure is simple and rapid. Injecting calcium solution in the vein or body cavity of the cow can bring about an amazing recovery. This, however, is something a veterinarian should do as improper administration can also kill the cow.

Milk fever is usually evidenced by a dullness of eye and lack of appetite. The cow feels cold to the touch. If you don't have an animal thermometer, place your hand under the rear flank near the udder. If the cow feels cold to the touch, her temperature is probably low.

In severe cases of milk fever, the cow goes down and actually becomes unconscious with the head pulled around to the side of the body. If you find a cow in this position, waste no time in calling the vet. The cow is in big trouble.

Often a cow can't get up for a while after a bout with milk fever. Some cows, if left alone, will respond in a few hours. Others have been known to lay for hours before getting on their feet again. Often another treatment of calcium helps the cow to respond.

Another common problem is the cow who misses the gate and goes through the fence, resulting in a painful cut on the teat. If serious, call the vet. For minor cuts, home remedies can be used if care is taken. Be sure the injury is clean. Often a good ointment and a band-aid will do the trick.

The major problem with a cut teat is the soreness and the cow's reaction to the hurt. Just be sure the sore quarter is always milked out. Don't give up out of disgust and leave milk in the injured quarter. This only leads to more soreness, more protesting, and perhaps even to a case of mastitis.

Basically, mastitis is an inflammation of the udder resulting from improper milking practices or disease, usually caused by strep or staph germs. Even an insect sting can cause a swollen quarter and bring on mastitis. The chunks that come from an infected quarter are a sign that the quarter is trying to heal itself. Mastitis caused by an outside injury such as a cut or bruise may be best left untreated.

Keep the injury clean and make sure the quarter is milked dry after each milking. If the mastitis doesn't clear up after a few days, call the vet. He will probably have the best answer.

If your stubborn cow refuses to milk by hand on the sore teat, a milk tube (available in the veterinarian departments of some pharmacies and farm supply stores) inserted carefully up into the teat canal will allow the milk to run freely from the sore teat. This takes the pressure, caused by milk, off the injury and relieves the soreness.

For the lone cow, heat is sometimes difficult to detect. Usually the animal becomes very nervous and often noisy. Milk production may slump during a heat period. Following heat, there is a small bloody discharge that indicates the cow has been in heat. If heat has not been detected before then, make a note on the calendar and watch the cow closely 20 to 21 days later for another heat period.

Casting of wethers can occur, but thankfully is rare. Generally happening within a short time following calving, the cow continues to strain and will cause a prolapse of the uterus, turning the organ inside out and pushing it out so that the organs are hanging outside the body. It is a horrible thing to witness and is often fatal.

If the veterinarian can't arrive soon enough, clean off the organs and push them back in place. The cow has a chance of survival this way. The broken artery is the greatest danger, as the animal can bleed to death internally.

Do not try to treat the cow alone. It takes an expert, often with your help, to put her back together.

Some cows will not breed again after casting wethers. Others will breed only to do the same thing the next calving. Some cows never have any trouble again. Perhaps the best answer is to milk the cow until she is dry, then sell her for beef or better, if you can eat a pet, put her into your own freezer.

Chapter 2

Horses

Probably no animal on the homestead places as much responsibility in your hands as work horses do (Fig. 2-1). Beyond ordinary care, they must be protected from work injuries, from each other, and from your own ignorance.

Men and horses don't change much. Horses always seem prone to skittishness and herd orientation, but they are still generally trusting after you have truly befriended them. The novice probably knows nothing but fear when starting with big horses and after the first drive will have learned enough to be dangerously overconfident.

ACQUIRING DRAFT ANIMALS

The first question to consider is whether or not you have enough work for horses. They can't spend all their time eating and standing around, so have at least 10 to 20 acres farming potential before investing. It is also desirable to have large woodlots and dirt roads available to nice places.

Think through your farming and other tasks. What crops, how much of each, how to plant, tend, irrigate, and harvest—all should be thought out in advance. The percentages of feed crops, human food, or cash crop to be grown must also be decided early. Will the team provide the year in, year out power to fully accomplish tasks that are started?

The price of this investment, as opposed to other power sources, must be considered. This balance is often settled in terms

Fig. 2-1. Work horses are invaluable on a homestead (photo by Jean Martin).

of the availability of necessary equipment. Horses and tractors are simply two forms of energy, varying in power and traction, that work through the tools attached to them. Consequently, the more functioning tools you have, the more kinds of tasks you can perform. Thus, if you want power for a big farm or ranch, and can afford the necessary implements and maintenance, get the tractor. If traction holds greater potential for your terrain and the tasks can be completed more slowly (because wet ground that can be traversed by horses stalls machines), then consider further the draft horse.

Finding and repairing old tools bought at low prices is a major gain. If you think you have the inclination for the hard work necessarily involved using draft horses, then proceed to find the men who can repair the equipment for you if you can't do it yourself.

The Draft Horse Journal, Route 3, Waverly, Iowa, is a good starting point for locating draft horse farms and ranches. Write many letters and visit many places. Ask questions of those who obviously know their trade.

There are many things to keep in mind when purchasing a draft animal. If the seller gets irritated when you ask questions, he is not the man to buy from. Does he work his horses on the farm, in the woods, show or breed them, or some combination of those? How does this relate to your homestead work? Is the seller honest about horses with deficiencies? If you can't get a vet check, what sort of guarantees will the seller make on the animal?

What hitches does the seller use? Is he anxious to show you the team at work or does he stall around? Ask about his fields and tools. Is his sole livelihood his team's ability to work? If not, how could yours be if buying those animals? Does the animal let you lift his hooves? Why is he selling the animal? Mix suspicion, curiosity, and questions and soak up all the answers.

We have yet seen anyone prove that one draft breed is better than another. Size, temperament, and match are truer considerations. Purebred horses come registered or grade (unregistered). Some horses are crosses (as half Belgian, half Percheron) or occasionally half bred to riding stock. An animal between 1400-1800 pounds is more than ample for most any homestead task. One-ton animals are not necessarily better because they are bigger. Donkeys and ponies can do many things, also. Animals sell according to their experience, which comes usually to animals of even temperaments. Conformation is relevant to breeding and general physical condition, which must be sound if the animal is to work for several more years. Matching is important in relation to length of stride and weight. Color is a consideration for show only.

Although prices vary greatly, young quality stock seems valued at around $500, greenbroke (which is little better than unbroke) at $600, and solid middle-aged teams at $750 or more per animal. Old teams, provided they aren't ready to kick over, can be excellent, mellow buys, as they sell for less, but the heart of an animal is more important than his hide.

Halter and collar should be provided at no cost for the above prices. Both must fit well, especially the collar. Harness that is old but in working condition shouldn't cost more than $100. Try to coerce a solid, uncracked pair of lines from the seller, as this is probably the most important single piece of harness, other than sized collars.

Another question to consider when chossing animals is their sex. Mares have obvious breeding potential, which on many ranches is the only function they perform. They are occasionally fussy though. Geldings are often more malleable to work patterns. Stallions are not something to start out on. Colts of either sex should be "mannered" by two years of age, started by three, and trained during the next four years of their eventual repertoire.

Lastly, transportation must be arranged to close the deal. It is preferable to get a stock truck, as a horse trailer is usually too small.

Do not solve feed and shelter problems after procuring the animals. You'll probably have to buy hay until you can grow enough of your own.

Alfalfa hay should not be fed to draft horses all the time as they are not being fattened for a dairy or meat product. Feed oat hay or any other grain or grass hay, as long as it is clean and full of seed. Some grain hays have whiskers that get too sharp if they dry out, so cut barley and wheat hay early. Oats, corn, and rye are good grains to feed for supplementing thin hay and after long work days. Include green grass when possible, but if you are in a working season, don't turn them out to pasture. Keep the horses available and out of trouble.

The barn should be large and solid. The team can withstand some weather, but they should be protected from the brunt of winds and precipitation. We prefer an adjacent wooden corral with a couple of extra-solid tie posts or, even better, tie stalls. Tying horses must become a regular and very conscious habit. Tie them high and short with just enough slack to feed properly (2-3 feet).

We've found the best routine to be tying them for feeding and then letting them loose in the corral when not needed for work. Leave halters on if there is any doubt in your mind about being able to catch them again.

These barnyard routines are strong elements in developing a rapport. When you feed, don't let the horses climb over you to get the hay. Make them "whoa" and keep the flakes away from their reach until they settle down. This keeps their attention and makes mealtime a learning situation. Keep water available at all times.

The only way you will know the condition of hooves is by picking them up and looking at them. Simple? Hopefully, but not necessarily.

Assuming you have doubts about picking up the foot of a reluctant 1-ton animal, let us review some attitudes towards animals in general. In a new situation, they are nervous; they usually respond to a calm, firm presence. Don't make them guess or worry about where you are or what you are doing. Be physically and vocally on their wavelength. Don't expect them to figure out yours.

When cutting a hoof, you may need a post or friend to hold the animal. Since the horse may lean, you must be well-braced against his shoulder. Push the weight off the hoof and lift by the shank of ankle hair that most big animals carry. Since the animal may paw or test or even just plain kick at you, it is obvious why the person who sells you the horse should go over it with you. Rasping and clipping are jobs for experienced hands. The only reason we would shoe a draft horse is because the animal will be working continually on pavement or to correct deficiencies in hoof structure. Be persistent, but calm, at all stages of handling and training.

We have not even broached the subject of actually working the animal because we believe that to continue to that stage, you must have safely arrived through this first stage. If you've acquired the right animal and don't kill it with love (and if you will positively not be turned aside by discouragement or hard work), keeping a short lead rope, a tight line, and using a kind voice will get you well into working with draft horses.

TRAINING A COLT

There is nothing more frustrating to a person than to raise a nice colt to the age it can be broken to ride and drive and then have it be unmanageable and unsafe for the family. How many colts bite, strike, kick, shy, or run over their owners? We're not talking about an occasional happening while playing, but regular occurrences. No colt, unless it's half dead, never plays or acts up. It is reasonable to expect a well-mannered, safe colt, no matter how high lived. Let's look at some of those habits and see just where they come from and what you can do to prevent them.

Biting

Most colts bite naturally in play. They race around with each other, taking a nip here, a nip there. So it is natural that they begin to nip "their" people, too. The first signs you will notice will be lipping at your sleeve, arm, or the back of your neck. At first it will just be a gentle rubbing of the lip back and forth, but if not stopped, it will grow into snatching, pinching, and finally really crunching down. Stop it before it really gets started. A good cuff of the neck will usually discourage it. If this is not enough, don't be afraid to smack the colt's face or nose. A lot of people say that makes a head shy colt. This is not true, unless you beat the face or continually slap the face. This is not necessary if you stop the nibbling at the beginning.

Never feed the colt anything out of your hand. This teaches him to nibble and bite, demanding goodies. When the horse is well-mannered, a horseman can reward him with edibles, but not the novice. We guarantee a biter when the novice continually feeds goodies from the hand.

Never play with the colt's nose (Fig. 2-2). People often rub the nose or around the nostrils and lips. This can encourage biting. Pet the colt's neck or jaw.

Fig. 2-2. Playing with a colt's nose can encourage biting (photo by Jean Martin).

Striking, Kicking, and Shying

These three vices all stem from two things: lack of exercise (of the right kind) and lack of discipline. An energetic colt will tear around, pretending he is frightened, bucking, kicking, and shying. This is normal and good. When he is being handled, he should act calm and well-mannered. The colt, in play, will kick and strike his mother, and will revert to playing with you in the same way.

Unfortunately, you are much smaller than his mother, and that hurts. By training and kind discipline, you must show him that such actions are not appreciated and will not be tolerated.

First, you should try to prevent these behaviors from happening in the first place by giving the colt plenty of good exercise, such as free exercise in a pasture with other colts his age, running along with his mother while she is being ridden, and, most important of all, training. Don't let anyone tell you that a young colt cannot receive training. This is not true. Training can begin at birth. Halter breaking, yielding all four feet, standing for brushing, work on the lunge line, and taking long walks are all work that will occupy the colt's mind, provide exercise, and improve his manners.

Should the colt ever kick or strike, correct him at once by using a shout or firm jerk on the lead rope. This is usually sufficient.

If the colt should shy, keep a firm hold on the lead while speaking to him quietly. Don't automatically think him really frightened and soothe him (some colts eat that up, and as a result, shy more often). Instead, try to see why he shied. If there is no reason, speak to him once, then ignore him.

Often shying (pretend shying) will extend into pushing or running you over. Here the colt will become excited and just shove against you, knocking you right down in some cases. This is not only annoying, but downright dangerous, especially with young children around. As with all bad habits, it is best to curb this before it gets established.

Correct handling is most important. When leading a colt, always grasp the halter with the right hand. Never hold the middle or end of the lead shank alone, as it gives the colt more maneuvering room (and he will use it). If you are holding the halter with your right hand, you can easily bring your right elbow up, poking the colt's neck or chest should he shove against you. Handled in this manner, few colts will get into the vice of pushing. With the wrist, you can tip his nose down, stopping forward motion, while the elbow discourages him from getting too close to you.

A good colt is a combination of two things: breeding and

training. In most cases, you have the control over both. So the old saying is very true: "A good horse is what you make him!"

HARNESSING A HORSE

At least one historian has noted that the invention of the horse collar, along with the divine profusion of water, light, and air, stands as one of those apparently insignificant things we take for granted without which we would surely perish.

Simple in design, the importance of the horse collar is matched only by its modesty, for only after its relatively recent development (France, tenth century thereabouts) was man able to put to the task of food production and transporation a force significantly greater than his own back. Certainly the Romans, Greeks, Assyrians, Babylonians, Egyptians, and the ancestors of Abraham in the Land of Ur had horses. Their horses pulled—not pushed—their loads and never was there such waste of motive effort.

Horses in those days were tied (or awkwardly yoked), not harnessed to their work. How much weight can a horse put on a choking breast snap, or a rope around the neck or saddle horn, or, worse still, a simple line from the load to the animal's tail?

Men had horses for thousands of years, but until they learned how to collar them, they failed to reap the full power at their command and nearly starved in the process.

Once the collar had found its place, however, horsepower rally started to mean something. With two good-collared horses, a man could plow enough ground, plant enough seed, cultivate enough crops, and harvest and haul enough foodstuffs to feed himself and his family all winter and have some left over for his friends in town. (This started a movement. Today, practically everybody lives in town.)

It is most important to start with the right collar. If it is too large, the horse will push himself through it and become "sweeneyed." (Their is undoubtedly a more accurate veterinary term to describe this situation, but you don't have to know how to spell sweeney to know how bad it is. A sweeneyed horse, suffice it to say, is soon fox meat.)

If the collar is too small, the horse will be short of wind. He can breathe all right standing still, but when he's working breathing isn't enough. He needs wind, which he can't get through a collar that is too small.

For the sake of expediency, we must assume you have the right collar, have the horse in his stall properly haltered, and the harness already hanging where it belongs. Enter the barn and speak nicely to the horse while approaching from the left rear. Do not approach abruptly and silently from directly behind. This may upset the horse, who will not view the harnessing procedure cheerfully under the best of circumstances and is probably just looking for an excuse to boot you out of the barn.

Entering the stall on the left side of the horse, brush him down a little. Push him over to his right with a mighty heave. He'll probably heave right back on you, but there is a cure for this, which we'll come to in a moment.

Some horses require a collar pad (a big, quilted thing that goes over the horse's neck). This complicates things somewhat. It can only be placed on the horse in one logical way, however, and is often omitted, so we'll spend no more time on this item.

Assuming the collar pad is not involved, you then find the proper collar and proceed to a most important step. With a dull knife, gently scrape the accumulated sweat, salt, dirt, and so forth off the collar where it touches the horse. This will prevent sores in the shoulder region, which is important, because a small sore on a shoulder can put a larger horse out of business for a long time.

You may have gathered, by this time, that horses are rather delicate creatures, and this is true. Proper harnessing is governed not only by the design needed to utilize the horse's strength to best advantage, but also by procedures that keep him from killing himself while he works. You cannot rely upon the horse to look out for his own body. In truth, it would appear that self-destruction is instinctive in a working horse.

Unbuckle the collar (collars are stored buckled to prevent breaking away in the wrong places). Do not attempt to put the horse's head through the collar like a noose. It won't work. Open the top of the collar and put the collar on from under the horse's neck, rebuckling at the top. The large end of the collar is at the bottom while the small end of the collar is at the top.

While you are reaching up doing this, the horse will, without doubt, attempt to step on your foot. Do not simply stand there and yell. Brace your hips against the wall and push the horse onto his far foot while you kick his instep with your free foot. Chances are he'll relax enough so you can tear free.

Go back and pick up the harness. There is a trick to this, regardless of whether you're left-handed or not. The harness is hung on two pegs. To the right is hung that portion of the harness that crosses the animal's rump. This varies with harnesses, but is

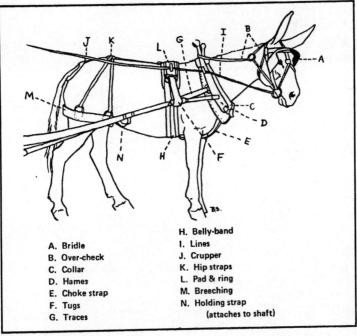

Fig. 2-3. Single wagon harness.

A. Bridle
B. Over-check
C. Collar
D. Hames
E. Choke strap
F. Tugs
G. Traces
H. Belly-band
I. Lines
J. Crupper
K. Hip straps
L. Pad & ring
M. Breeching
N. Holding strap
   (attaches to shaft)

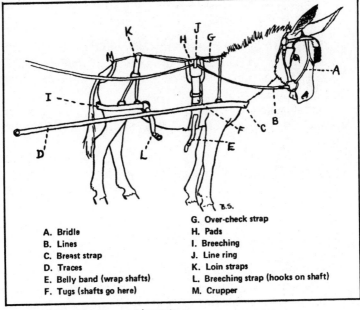

Fig. 2-4. Basic American cart harness.

A. Bridle
B. Lines
C. Breast strap
D. Traces
E. Belly band (wrap shafts)
F. Tugs (shafts go here)
G. Over-check strap
H. Pads
I. Breeching
J. Line ring
K. Loin straps
L. Breeching strap (hooks on shaft)
M. Crupper

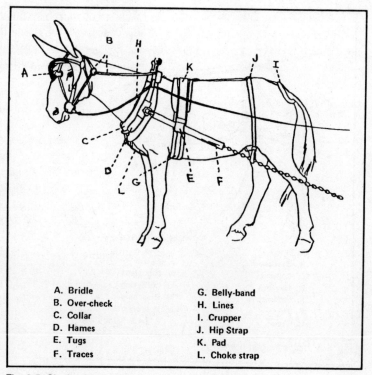

A. Bridle
B. Over-check
C. Collar
D. Hames
E. Tugs
F. Traces
G. Belly-band
H. Lines
I. Crupper
J. Hip Strap
K. Pad
L. Choke strap

Fig. 2-5. Single plow harness.

usually a good big piece of leather to which are attached (when not in use) the tugs, the hangers for the breeching strap (pronounced "britchin"), in addition to straps emanating from this point all over the horse. See Figs. 2-3 through 2-5.

Put this piece on your right shoulder. This leaves both hands free to grasp the hames. You will recognize these because there is hardly anything else to grasp.

Thus burdened, get in beside the horse again. Place the hame in your right hand over the horse and down along the collar on the right side. Bring the hame in your left hand tight against the collar on the left side.

Heave the rest of the harness over the horse's back, transferring the load from your right shoulder to its proper place on the horse's rump. If the harness was put away properly, it will fall naturally into place around and over the horse. If the harness was put away in a mess, you might just as well quit because nobody could tell you in one lesson how to straighten all of it out.

This business is not easy because, while you were picking up the harness, the horse has shook his head and maneuvered the collar you fixed so nicely up on his neck to a point just behind the ears. You can shove the collar back where it belongs, however, without letting loose of the hames.

Fasten the hames under the collar, making certain they're positioned properly in their grooves. On most hames there is a sort of clip at the bottom giving you leverage. The hames have to be extremely tight, or they will pull off the collar and injure the horse.

There is a large strap that passes under the horse's neck, fore and aft. Reach down and pass this between his front legs and up under his belly. Grope around under the horse until you find the end of the belly strap that goes under the horse's middle and under the strap coming from the hames. Snap or buckle this belly strap to the other end of this strap on your side of the horse.

While you are down under the horse finding these straps, he will lean over. When you try to stand up, you will find there is no room left for you in the stall between the horse and the wall. To fix this, get a piece of broomstick about 2 feet long and hold one end against the wall when you stoop over.

When the horse leans heavily against the other end of this stick, he will nicker and quiver a little. He will voluntarily get back over where he belongs so you can go on with your work.

There are two more straps, leading generally from the ends of the breeching on either side. Grope around until you find these and snap them into the ring on the big fore and aft strap held up by the belly strap.

When these steps have been completed, you will notice the horse's tail is under the breeching. He looks very uncomfortable. Pull out the breeching and lift his tail out, taking care to duck quickly to one side when the horse gratefully whips his tail in your face.

With some minor adjustments and shifting around, your horse is now essentially harnessed. Go and get the bridle.

Do not unhalter the horse to put on the bridle or he will gleefully leave the barn, harness and all, leaving no forwarding address.

Get the bit all set to insert, put your right thumb in the horse's mouth, high enough up along his cheek to avoid his greedy incisors, open his mouth, and place the bit on top of his tongue. If the bit is under his tongue, he will commit suicide.

Then work the rest of the bridle up under the halter nose strap and over the horse's head, putting the chin strap under the halter at all points, and buckle same. The halter can then be removed.

Back out the horse and lead him outside where the lines, neatly coiled on the left hame, can be laced through their proper rings and attached to the bit.

Your horse is now harnessed and you've passed the test but, of course, in most instances you would harness more than one horse to get anything done. In such a case, you have two horses to lead out of the barn and only need to know a couple of tricks to hitch them up to your cultivator.

One trick helps solve an age-old problem that is created when two horses accustomed to working side by side leave the barn with the wrong horse on the wrong side.

You can remedy this, not by asking one horse to stand still while you lead the other one all the way around him (which doesn't work), but simply by holding both horses tightly by their bit rings, raising their heads, and walking briskly down between them turning them neatly inside out as it were. This is a maneuver no horse has yet been able to figure out, and it works admirably.

Once your team is properly rigged, get behind them and drive them up to your machine from the side. Encourage one horse to step over, not on, the tongue, and knuckle them into position with a minimum of backing and filling.

Always fasten the neck yoke first, thus lifting the tongue and avoiding the unpleasantness of a runaway with the tongue bobbing along on the ground. Hook up the tugs to the singletrees, experimenting until you discover which link is right but always using the same number of links on each side of each horse. Trees should be as close as possible and still just clear the horse's hocks in full stride. Hang your water jug on the nearest hame, mount to the seat, and away.

HOOF CARE

The value of a horse lies chiefly in his ability to move; therefore, good feet and legs are necessary. The foot structures consist of bone, cartilage, ligaments, tendons, fatty tissue, highly sensitive flesh, horn, blood vessels, and nerves.

One of the first and foremost things in hoof care is daily cleaning, which will prevent most ailments and lamenesses or be able to correct such before it gets too far along. Secondly, handling the feet often helps the horseshoer (*farrier*). Remember, the better the horse stands, the better job the horseshoer will do.

To properly pick up the horse's front foot, place the left hand on

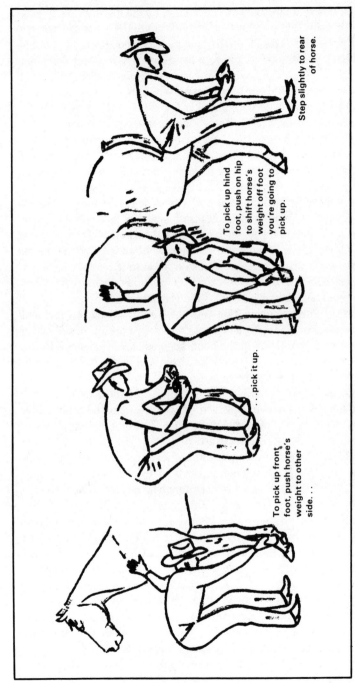

Fig. 2-6. Steps for picking up a horse's front and hind feet.

the horse's shoulder and push his weight off the leg you wish to pick up (Fig. 2-6). Run your hand down the front of the leg to the pastern and lift the foot. To pick up the hind foot, push on his hip to put his weight on the other side, run your hand down the leg, and pick it up. If he should try to kick while you are picking up his foot, it is less likely to connect with you when you pick it up properly.

Cleaning is done with a tool called a hoof pick, which can be purchased at your local feed store for about a dollar. A bent screwdriver can also be used, but the main thing to remember is daily cleaning of the feet. Scrape the mud and material from the heel toward the toe of the hoof. It does not hurt the horse. Be sure to search carefully for embedded foreign objects such as a nail that may puncture the bottom of the foot and will cause the animal to go lame. While you are cleaning the feet, also inspect for loose shoes. Make sure, also, that your horse has an annual tetanus shot.

One of the most common problems with horse's feet is *thrush* (hoof rot), which is one of the easiest to cure if you will devote a little time and effort to eradicate it. Thrush is caused by barnyard moistures and the lack of air to the tissues of the frog.

This starts a fungus that rots the tissue, giving it a foul and nauseous odor. If the frog tissues continue to decay, it will cause more severe problems in the hoof, such as lack of pressure to the plantar cushion (Fig. 2-7), which is located directly above the frog and circulates the blood throughout the hoof. This would mean faulty circulation of the blood flow. Also, without good frog pressure, the foot may suffer contracted heels, meaning the heels would grow too narrow in the back of the hoof. The frog normally causes pressure to push those heels apart.

Thrush is commonly seen in horses with poor hoof care and those kept in wet, dirty environments. The cure is relatively simple: daily cleaning and the use of common household bleach (about a capful) around the frog, letting it seep down into the tissues. The bleach solution can be straight from the jug and should be enough to soak into the tissues. This is to be repeated faithfully until the organisms are killed. There are products on the market for this purpose. Kopertox is excellent, but we have found common household bleach readily available, very effective, and inexpensive.

The problems connected with the care of the horse's feet are becoming more numerous every day as we continue to change the natural life-style of the horse. A good farrier can be of tremendous benefit to your horse and to you as a horseman.

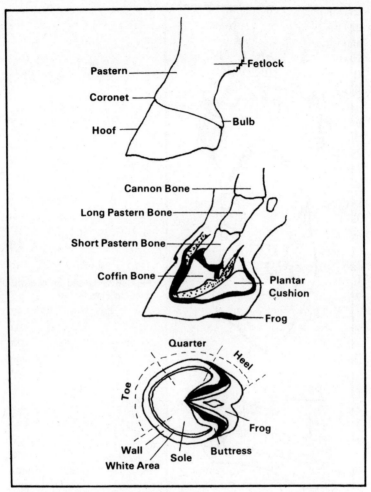

Fig. 2-7. Parts of a horse's foot and hoof.

TEETH

To the novice horse owner, there's probably nothing so ugly looking about a horse as his teeth. To look at the horse and brush a hand across a lightning-fast coat, over a fat rump, down sleek legs, and stop, in praise of a lovely face, who would guess that behind that teddy bear muzzle are teeth that Count Dracula wouldn't be caught dead with? Where have they been hiding the Ultra-Brite?

A horse keeps his mouth shut most of the time and, despite the frightful appearance of a horse's teeth, they do seem to work. In

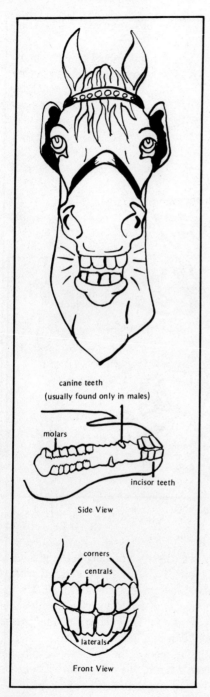

Fig. 2-8. Side and front views of a horse's jaw.

fact, besides being functional, they can supply us with information about his age.

It seems strange to me that more horse owner's do not know how to tell a horse's age by his teeth. Although a not altogether accurate measure, a horse with no registration papers has a birthday subject to the whims of his owner. To the buyer, though, an approximation, of the horse's age is a must.

The first step in reading a horse's teeth is to know where they are situated in the skull. According to the positions in the jaw (Fig. 2-8), the teeth are classified as *incisor teeth, canine teeth* (usually found only in males), and *molars.*

In telling the age of a horse, the only teeth we need look at are the incisors. The incisors (Fig. 2-9) are split into three pairs—the *centrals*, the *laterals,* and the *corners.*

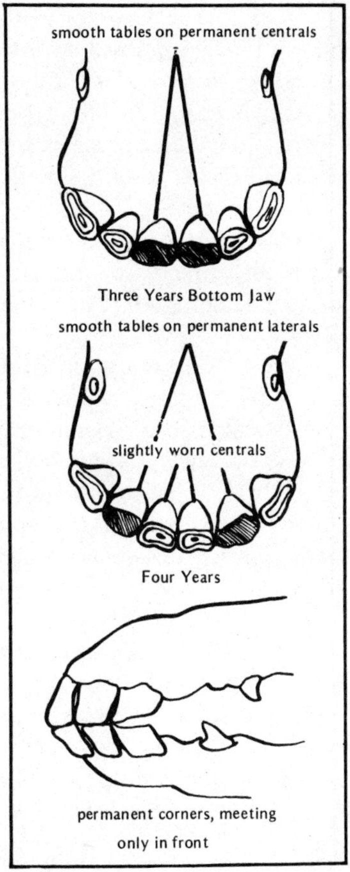

Fig. 2-9. A horse's teeth when he is three, four, and five years of age.

smooth tables on permanent centrals

Three Years Bottom Jaw

smooth tables on permanent laterals

slightly worn centrals

Four Years

permanent corners, meeting only in front

A young horse's first teeth are called *milk teeth*. These are small and very white. The horse has his milk teeth from when he is a few weeks old until he is three years old.

At three years of age (Fig. 2-9) the horse will lose his central milk teeth. These will be replaced with permanent teeth. The permanent tooth is larger than the milk tooth and more yellowed. The tops, or tables, of these teeth will have no markings on them yet because they have not worn.

At four years of age (Fig. 2-9) the horse will get his permanent laterals. They will look the same as the permanent centrals did. By this time however, the tables of the central will have worn down and slightly darker rings will show in them. These are called cups.

At five years of age (Fig. 2-9) the permanent corner teeth come in. Again the tables are flat, but when the teeth are seen from the side, the top and bottom of the corner teeth meet only in the front.

At six years the corner teeth will have worn level and all the tables of the incisors will be in wear—have cups. At seven years (Fig. 2-10) a hook develops on the back of the upper corner incisors.

At eight years (Fig. 2-10) the hook is worn away and a dark line appears in the tables of the centrals. At nine years there is a dark line on the tables of all the teeth.

At 10 years (Fig. 2-10) the central teeth become more triangular in shape and the cups become fainter. Also at 10 years the slope of the teeth has increased, and a dark groove will appear at the top of the upper corner tooth. This is known as the *Galvayne's groove* (Fig. 2-10).

After 10 years it is much harder to determine the specific age of a horse. With increasing age the slant of the teeth becomes more acute, the tables become more triangular, and the cups disappear. Also, the Galvayne's groove will get longer. At 15 years it will be approximately halfway down the tooth. At 20 to 25 years the groove will start disappearing at the top of the tooth. Beyond 25 years it probably doesn't make much difference how old the horse is—he's just old.

This is only a fairly accurate way of telling how old a horse is. It's really more of a way to get an approximation of age. Horses' teeth wear differently depending on their food and eating habits. A pasture grazed horse will look a year or so older by his teeth than a stall fed horse because he wears the teeth more. The same may be true of a horse who 'cribs' or chews on wood. It is a handy thing to know, and it may keep a horse dealer from pulling a fast one on you.

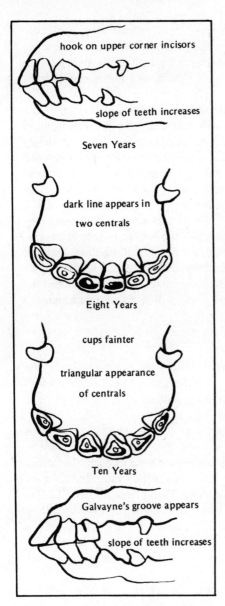

Fig. 2-10. A horse's teeth when he is seven, eight, and ten years of age. Note the Galvayne's groove.

Now to the care of your horse's teeth. He may need a little minor dentistry now and again.

Horses have *wolf teeth*, which are a pair of small, virtually rootless teeth in front of the molars. Often times they can be a bother to a horse as the bit could bang against them and cause soreness. If your horse seems fussy with his head and mouth when you are riding him, there's a possibility wolf teeth could be your problem. A vet can easily remove them, and the charge is fairly minimal.

A very commonplace dental problem in horses is rough, uneven molar faces. In your horse's yearly checkup it is important to have your vet check his teeth. If the edges are rough, he will simply file them down. This is called *floating* teeth. Rough molars left unattended could cause serious problems to the health of your horse. Lacerations of the mouth and tongue can occur and weight loss is usually apparent as the horse cannot chew properly. If you have purchased an underweight horse, check to see if his teeth may have led to the condition. Gently pull the horse's tongue to the side and feel the molars for sharp edges. A simple unchecked problem like this can lead to a lot of worrying and vet bills. The daying "no foot, no horse" could just as easily be no teeth, no horse. Keep the teeth healthy and the rest of the horse will be a whole lot happier to do the same.

ENERGY AND FEEDS

Energy can be defined as the ability to do work, which can then be changed into useful functions for the horse. This can be the horse simply maintaining itself while doing nothing more than standing in a stall or gazing in a pasture.

Work is required for growth and for the actual physical labor the horse performs. There is work involved when the mare is creating a foal gestation and for her to produce milk during lactation.

Energy can be measured by various units. Units often used are calories, kilocalories, and total digestible nutrients (TDN). TDN is equal to the sum of the digestible protein, digestible carbohydrates, and the digestible fat multiplied by a factor of 2.25. The digestible fat is multiplied by a correction factor of 2.25 because fat contains more energy than proteins and carbohydrates. We will measure energy here in terms of TDN.

The amount of energy required by the horse is dependent on the energy output of the horse. A horse working hard requires more energy (or pounds of TDN) than a horse that is not being worked. This is common sense. A hard working horse is fed more grain and is supplied with more energy. Table 2-1 gives the energy values necessary for a horse maintaining itself and the additional energy requirements for growth, work, gestation, and lactation. Protein

Table 2-1. Energy Requirements for Horses for Different Functions.

Function	Energy	
	TDN (lb.)	Dig. Energy (Kcal)
Maintenance, per 100 lb. body-weight (B.W.)	.80	1600
Growth, per lb. gain (above maintenance)	1.80	3600
Fattening, per lb. gain (above maintenance)	3.85	7700
Work*, per hour per 100 lb. B.W. (above maintenance)		
Light	.12	230
Medium	.29	570
Heavy	.55	1090
Gestation, per day (above maintenance) for last 4 months	.50	965
Lactation, per lb. milk (above maintenance)	.18	360

Energy requirement for work depends on the weight of the horse and the degree of work expended.

requirements must be met also; however, we will deal only with energy needs here.

From Table 2-1 the energy requirements of an average 1000-pound horse can be calculated. If on the average 0.8 pounds of TDN are required per 100 pounds, then a 1000-pound horse requires 8.0 pounds of TDN per day, as illustrated below.

$$\frac{0.8 \text{ pounds TDN}}{100 \text{ pounds horse}} \times 1000\text{-pound horse} = 8.0 \text{ pounds of TDN}$$

Additional energy is required for a horse doing work. A 1000-pound horse doing light work requires 9.2 pounds of TDN daily.

$$8.0 \text{ pounds TDN} + \frac{0.12 \text{ pounds TDN}}{1000 \text{ pounds horse}} \times 1000 \text{ pound horse} = 9.2 \text{ pounds TDN (total energy) required for light work}$$
(for maintenance) (additional energy required for light work)

Following the same procedure, the requirement for a 1000-

Table 2-2. TDN Requirements for a 1000-Pound Horse.

Maintenance	8.0 lbs.
Doing Light Work	9.2 lbs.
Doing Medium Work	10.9 lbs.
Doing Heavy Work	13.5 lbs.
Mare Last 4 Months of Gestation (no additional work)	8.5 lbs.
Lactating Mare	15.9 lbs.

pound horse doing medium work is equal to 10.9 pounds of TDN. This same horse doing heavy work needs 13.5 pounds of TDN. The broodmare requires additional energy during the last four months prior to foaling. A 1000-pound broodmare requires 8.5 pounds of TDN each day in the last four months of gestation. A comparison between the broodmare and the horse doing light work shows the broodmare does not require as much energy as a horse doing light work. She does require more energy than the horse simply maintaining itself. Once the mare drops her foal, her energy needs increase greatly.

During the peak of lactation, her energy requirement increases to almost 16 pounds of TDN or twice her maintenance requirements. Table 2-2 summarizes the requirements for a 100-pound horse. These are just average values. The proper feeding of a horse is the major role of the "manager" and falls more into the realm of art than science.

Energy is also required for growth—1.80 pounds of TDN for each pound of gain. When the foal is young and growing at a rapid rate, it needs more energy per 100 pounds than when it has almost reached maturity and is growing at a slower rate.

The energy the horse requires comes from feeds. Different feeds have different levels of energy. This energy is measured in terms of TDN (Table 2-3). In looking at this table it is obvious that TDN is not a perfect measure of energy. The TDN value for alfalfa hay is the same as the value for legume grass hay or an alfalfa grass hay. It is clear that the grass in the hay is of lesser nutritional value than the alfalfa; however, the value in the table is the same. Though the TDN method of measurement overvalues the energy available in poor quality forages, it is a method of general use.

Table 2-3 offers a wide range of alternatives for meeting the horse's energy requirements. Two factors are important. First, can the horse consume enough of the feed to meet the energy needs? On paper it might look like straw could meet the energy requirement; however, the number of pounds of straw the horse would have to consume exceeds the amount a horse could eat in a day.

The second factor is the economy and availability of various feeds. As feed prices increase, most concerned horsemen will want to meet the energy needs of their horses for the least cost. The situation in many areas this past winter was such that hay was relatively expensive and corn cheap. The energy requirement for the maintenance of a 1000-pound mature horse is in the average 8.0 pounds of TDN per day. The price of mixed hay varies, but in the Midwest it sold for approximately $80 per ton. To meet a horse's energy requirements with mixed hay would cost 64 cents per day using the energy values from Table 2-3 and the estimated price in the following calculations.

$$\frac{8 \text{ pounds TDN required}}{0.5 \text{ pounds TDN pound of hay}} = 16 \text{ pounds of hay required}$$

$$16 \text{ pounds of hay required} \times \frac{\$80}{2000 \text{ pounds of hay}} = 64 \text{ cents}$$

To meet the energy requirements of the same horse with oats will cost 73 cents per day. To obtain this value the same procedures were used only using a cost value of $5.90 per hundredweight and a TDN value of 0.65. Using corn, the same requirements can be met for 45 cents per day. This value was arrived at by taking the cost of corn at $2.50 per bushel (1 bushel corn = 56 pounds) and the TDN value to be 0.8 pounds per pound of corn. If the energy requirements

Table 2-3. Energy Content of Selected Feeds.

Feed	Energy, per pound of feed (TDN, lb.)
Oats	.65
Corn	.80
Soybean Meal	.75
Linseed Meal	.70
Wheat Bran	.60
Molasses	.55
Alfalfa, dehydrated	.50
Alfalfa, hay	.50
Timothy hay	.45
Prairie hay	.45
Legume grass hay	.50

Fig. 2-4. Information on Oats and Shelled Corn.

Grain	Weight per Qt.	% TDN
Oats	1.1 lb.	65%
Shelled Corn	1.75 lb.	80%

of the horse can be met for less cost with corn last winter, why then do most horse owners feed grass hay and oats to their horses? Corn has a bad reputation for making horses fat or in some cases foundering them. Corn does cause problems, but this occurs when an oat ration is substituted with an equal volume of a corn ration. Many horse owners feed grain in terms of volume, such as a full coffee can. This practice in respect to oats and corn causes an essential doubling of the horse's energy intake. Table 2-4 shows that corn not only weighs more per unit volume but also has a higher energy value. The problems in feeding a corn ration are due to an increased energy value. The problems in feeding a corn ration are due to an increased energy intake when the horse is fed by volume rather than weight. This discussion is not to encourage everyone to feed their horses rations high in corn, but rather to emphasize the role of economics is shaping the horse ration.

The mature horse has energy needs that can be measured in terms of TDN. These needs are increased by growth work, and pregnancy. Various feeds have different energy values. We should try to meet our horses' energy needs for the least cost. In doing so, feed should always be measured by weight, not volume.

Chapter 3

Swine

Pigs are truly a much maligned and misunderstood animal. If you let them they will keep themselves relatively clean. They'll wallow in the mud if it's available, but if you keep them on an enclosed concrete surface, they behave very differently.

Pigs are the only farm animal who will sweep their own pens. They do this with their snout, pushing the manure off to one side of the pen.

Pigs do not thrive alone. They need competition at mealtime in order to eat properly. Fighting over the feed is the highlight of a pig's day. We fully understand the origins of the expression, "so-and-so eats like a pig." It's one of the few popular beliefs about pigs that is valid. They are very sloppy eaters.

Most people don't know pigs are experts in the art of screaming. They could easily be taped for the sound track of horror movies. Screaming is best observed at mealtime.

There are a number of very good, practical reasons for the small part-time farmer to choose hogs. They need no pasture, so large acreage is unnecessary. Although they will eat hay and can be pastured, hay is not a large component of their diet.

Corn is the mainstay of a hog's diet. We grow our own on our 40 acres and have it ground at a local feed mill. Commercial hog feed can be bought by the ton, delivered, or in 100-pound sacks. We store ours in cleaned-out 50-gallon oil drums we managed to obtain for nothing.

Space requirements for hogs are small. One needs not invest in fancy or expensive fencing to confine them. A single strand of electric fence is all we use. This works in both the sow pen and the pens where we feed out the young pigs.

There have, of course, been memorable occasions when the power went off and we've spent hours chasing pigs around our back pasture. As long as the power is there, the electric fence works.

Our total fencing needs for the 100 or so pigs we raise a year amount to about 550 feet of fence. The cost is under $25. In addition, we purchased a good electric fence charger for $35, making our total investment less than you would pay to fence in two cows.

There was an existing barn on our farm when we bought it, and we have converted part of the interior into pig pens. We use them in the late fall to feed out the September litter of pigs. It is usually too cold to use the outdoor pens. The indoor pens are made of scrap lumber; the only cost was for nails.

Indoor pens are not a necessity, so long as some shelter is provided in severe weather. We also have a small interior area walled off from the rest of the barn for the sows and boar. They can enter it at will from their outside pen.

Equipment needs are fairly minimal. You can invest in automatic waters at approximately $20 each plus some pipe, or you can buy several heavy-duty rubber buckets and pans and haul water from the nearest spigot.

If you raise a volume of hogs and keep them in indoor pens, a manure spreader and tractor would be necessary. For the indoor-outdoor type of arrangement that we have, we can get along fine with a large wheelbarrow and pitchfork. It's hard work, but it's cheaper.

Pigs will usually manure outside and leave their interior quarters relatively clean except in severe weather. We generally have to clean pens once a week in winter and not at all in the summer.

If you breed your own sows, your expenses will be somewhat higher. You will pay anywhere from $125 and up for sows and probably $200 for a boar. You could raise your own sows by starting with young piglets bought at six weeks of age for $15 to $30. In general, however, you're better off to start with an animal that already has a proven record.

For breeding purposes you will need either farrowing crates or small box stalls for the sows to farrow (give birth) in. New farrowing crates cost about $80 apiece. Box stalls can be constructed out of ½-inch plywood and two-by-fours.

We prefer the crates because they confine the sow more closely, and she has less opportunity to lie on and kill her young. We built our crates out of old pipe, using a pipe cutter and welder. The only cost we incurred was for the automatic waterers which we installed in each crate.

At first we had water pans in each crate for the sow to drink from. Several of the piglets drowned in the water pans. We switched to automatic waterers. Experience is sometimes a painful teacher.

An advantage to hogs is the amount of time to raise them from birth to market weight: only six months. This is a much shorter time than for beef, which usually require about two years.

SELF-FEEDER FOR HOGS

A self-feeder for hogs requires less labor than the hand-feeding system. Besides wasting less feed, the pigs have free choice.

The width of a 1 by 6 presently measures about 5½ inches. This design is based on a 1 by 6 being 5¼ inches wide, but the slight difference should have no major bearing on the end result (Table 3-1).

The skids may be cut from a 14-foot 2 by 4 which will project out at the ends of the feeder. If you bore holes through the skids near the ends, you can hitch the feeder to a tractor or team of horses and drag it to the desired location.

Place the partitions (Fig. 3-1) where you wish, depending on the variety and amounts of feed required. Place the roofing over the hinges. Put a separate piece over each door. The piece at top should

Table 3-1. Materials List for the Hog Feeder.

Lumber			
No. of pieces	Length	Dimensions	Use
2	12'	2" x 10" matched	flooring
1	12'	2" x 4"	skids
2	10'	2" x 4"	rafters and studs
1	12'	2" x 4"	triangular strips in corners of trough
15	12'	1" x 6" matched	flooring (actual measurements $^{13}/_{16}$" x 5¼")
5	14'	1" x 6" matched	flooring for roof
2	12'	1" x 6"	ridgeboard, side and ends
2	12'	1" x 6"	slides, triangular blocks, guides for slides, cleats for door
Lumber for desired cross partitions			
62 sq. ft. roofing paper			
Hardware			
6 heavy strap hinges		2½ pounds 6d nails	
1 pound 10d nails		four 2½" bolts with thumb nuts	

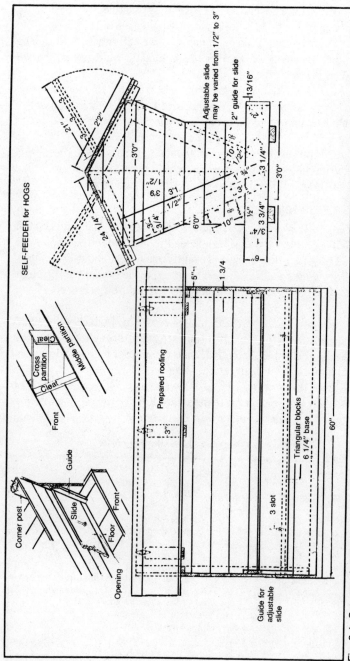

Fig. 3-1. Construction details for the hog feeder.

lap down onto the door about 1½ to 2 inches. The adjustable slides allow you to vary the size of the opening from ½ to 3 inches to accommodate different feeds.

HOG FEED

Under modern conditions and according to present theory, formulating hog feed is extremely complicated. It requires a knowledge of the nutrient levels of the various feeds available, as well as the nutrient needs of different classes of stock under different conditions. It involves use of such additives as antibiotics, arsenicals, nitrofurans, and sulfonamides. Mixing such feed requires a great deal of technical knowledge and the use of weighing and mixing equipment few hog producers can afford. Since very small amounts of some ingredients must be accurately weighed and mixed with very large amounts of other ingredients, the process requires delicate balances that are accurate to 0.1 grams (as well as other weighing devices) and a mixer that will evenly distribute minute quantities of vitamins, etc., throughout large batches of feeds.

Even very large hog producers are not able to economically justify the purchase of separate ingredients for feed supplements, and some of them (certain vitamins) do not stand up well in storage. Only large feed manufacturers can economically supply these ingredients.

Crystalline vitamins and drugs (called micro-ingredients) must be carefully diluted before mixing with other ingredients, and this, too, requires special equipment and skill. Other ingredients also require special equipment. One example would be waste animal fat, which entails a heated storage tank, metering pump, and blender.

Some hog feeds (especially pig starters) are in pelleted form for reasons of economics that are important to commercial hog farmers. Pelleting requires not only all the mixing and weighing equipment; it also requires a pelleting machine, which makes pellets from ground feed by steaming it and extruding it through dies. There may be as many as 20 different ingredients in a modern complete hog ration—without the grain.

If hog feeding is this complicated, why doesn't the feeder have to have advanced degrees in nutrition, physiology, chemistry, math, and a few other sciences? Because no one could have all that knowledge, spend all that money on equipment, spend all that time formulating and mixing feeds, and still have time to raise pigs. The farmer pays other experts to perform these tasks for him, and consequently if he takes their word for it, he doesn't have to know very much about it at all.

In contrast, the homesteader doesn't make use of these experts if he follows homestead feeding techniques. His methods fly in the face of advice from even the generalized hog experts. Any homesteader who seriously intends to compete with modern commercial hog raising methods probably has a lot of studying to do, and even the casual homestead hog raiser had best learn at least some basic facts about hog nutrition.

Practical homestead hog feeding, then, combines the best of modern scientific knowledge and some of the basic practices in the old-fashioned "scoop of this and scoop of that" method. One relatively simple way of looking at this (without going into any great detail on hog nutrition) is to examine what modern technology tells us hogs need and how agribusiness fills those needs, and then finding out how we can meet those needs on a homestead basis.

In the abstract, nutrition is a very simple science. All you need do is eat foods that contain proper amounts of the elements needed to sustain life and growth. What are those elements, what foods contain them and in what quantities, and how do they interact? In practice, nutrition obviously is not a simple science.

The homestead hog raiser is in a particularly difficult position because he often makes use of feeds just because they are available or cheap, because these ingredients change over the growing season, and because there is no modern research on what we're calling the homestead method of hog feeding. Anyone involved in any form of animal husbandry really should have at least a basic knowledge of nutrition, but for those who stray very far from bagged feeds with neat labels advertising protein, fat, fiber, and other important components, the need is even greater. You are what you eat, and the same holds true for a pig.

Food has two basic purposes. The first is obviously to keep the animal or person alive. It takes energy—fuel—to pump blood, to breathe, to digest, to keep warm, and even to blink eyelids and wag tails. Food is fuel, and there are high-grade and low-grade fuels.

The second purpose of food is to provide health and growth. This is especially important for young, rapidly growing animals, for pregnant females, and for females nursing young.

The life-sustaining aspect is most important, because if animals don't have enough nutriments to stay alive, there obviously isn't any opportunity to bother with growth or reproduction. The only reason for feeding hogs is to have them reproduce and grow, so we're concerned with a level of feeding that will do much more than just keep them alive.

Pigs need more than 35 individual identifiable nutrients. The amounts and proportions of these vary with their ages and the demands (growth, reproduction, etc.) placed on them. There is no single source for all these nutrients.

The two most important elements of feed, because they are essential to vital life processes, are energy and protein.

Calories

Energy nutrients are primarily carbohydrates and fats, although protein can also supply energy. Energy nutrients in excess of those required for vital bodily functions are stored as body fat. Energy therefore takes on added importance in the latter stages of hog raising as we "finish" the animals.

The energy value of a feed is commonly expressed in feed tables at TDN, or total digestible nutrients. Other nutritionists use DE or digestible energy. Perhaps the best way for modern city-bred, first-time hog raisers to think of it is in terms of calories. We're all familiar with calories in our own diets, and a calorie is a unit of heat, which is energy.

Plants contain different forms of carbohydrates. On feed tags these are listed as nitrogen-free-extract (NFE) or crude fiber. NFE includes sugars, starches, and some hemicelluloses—the more soluble carbohydrates. Crude fiber is cellulose and very complex carbohydrates. Crude fiber can be utilized by ruminants such as cows and goats because of the fermentation-vat function of their four stomachs. Pigs, like people, are monogastric and do not digest crude fiber.

On the average, cereal grains contain from 60 to 70 percent NFE and are low in crude fiber, which is why they are widely used in swine feeding. TDN values are arrived at by adding up all the organic digestible nutrients-protein, fiber, NFE, and fat-times 2.25. (Fat is considered to have 2.25 times as much energy value as protein or carbohydrates.) Multiply TDN by 2000 to convert to kilocalories of DE. Corn containing 80 percent TDN has a DE value of 1600 kilocalories per pound.

Grains are the chief source of energy for swine. While several other grains may be used, they are measured up against corn. Corn is deficient in many important nutrients. It is deficient in phosphorus, calcium, salt, vitamin A, vitamin D, B vitamins (including B_{12}, riboflavin, pantothenic acid, niacin, and choline), and the unidentified factors. It also doesn't have enough protein, and the protein is not of acceptable quality.

Protein is very important not only because it is the body-building ingredient in feed but also because it plays an important role in the body regulators such as hormones and enzymes.

The recommended protein content of feed for hogs ranges from 20 percent for creep feeding and 22 percent for early weaned pigs, decreasing to 18 percent at 40 pounds; 14 percent at 80 pounds; and 13 percent at 120 pounds. Gestating sows and gilts need 14 percent and lactating sows need 15 percent protein. Corn might have nine percent, but it is deficient in certain amino acids.

A feed can contain sufficient protein and still be deficient. Protein is a complex nutrient composed of amino acids, and the amino acids are what's important. There are at least 24 amino acids, but since they combine like letters of the alphabet, there could be as many proteins as there are words in the dictionary. Essential amino acids are those required by an animal but not synthesized. They must therefore be included in the ration. Looking at protein, per se, does not guarantee the presence of any given amino acid. Different animals need different amino acids, and they might consume a feed that is high in protein but deficient in the specific amino acids they require. In that case the "quality of protein" is poor.

Amino acids are needed for the formation of every new cell, so quality of protein is a basic requirement for growing pigs, and especially for young pigs. It affects not only meat (muscle) but internal organs, blood, and bone. Without the proper amino acids, the animals cannot develop properly.

When the swine ration consists of grain, which is deficient in crude protein as well as amino acids, the main problem of balancing a ration centers around correcting the amino acid deficiencies with protein supplements. It is generally acknowledged that the hog diet should consist of protein derived from both plant and animal sources, so common protein supplements include soybean oil meal and tankage, for example. On the homestead, milk and milk by-products may be the most important protein sources.

The most common supplement used to increase the protein level of corn-based rations for swine is soybean oil meal, which is 44 percent protein. (Because the oil meal is processed, it does not cause soft pork like raw or cooked soybeans do.)

The homesteader can buy soybean oil meal and still have organic pork. If the goal is complete self-sufficiency, homestead protein sources include milk (1 gallon a day), legume forage or rape (which is not a legume but which equals alfalfa in protein content for hog feeding), and other grains which are higher in protein than corn is.

The following amino acid requirements have been determined for weanling pigs, expressed in percent of total ration:

L-Arginine	0.20
L-Histidine	0.20
L-Isoleucine	0.70
L-Leucine	0.60
L-Lysine	1.00
DL-Methionine	0.22
DL-Phenylalanine	0.46
L-Threonine	0.40
DL-Tryptophan	0.20
L. Valine	0.40

Soybean meal and fish meal are good sources of these essential amino acids. Homesteaders can be assured that their feeding practices and scientifically sound too, by remembering Morrison's statement: "There is not apt to be a deficiency of any amino acid in a ration where the protein comes from three or more good sources, and where a considerable part is supplied by such feeds as soybean oil meal, fish meal, or dairy by-products. This is especially true when alfalfa hay or alfalfa meal is included in rations for pigs not on pasture."

In other words, with a feeding program that includes milk, vegetables, and pasture, in addition to grain, amino acid deficiencies are unlikely to occur.

Low levels of protein tend to produce more fat and less lean in the carcass, which is undesirable. Low protein levels result in slow and expensive gains. Conversely, protein is expensive and high levels of protein are simply wasted. Without souped-up protein concentrates, the homesteader is more likely to err on the low side and therefore should be aware of protein requirements and levels.

The safest course involves giving pigs access to *good* pasture or feeding legume forages, allowing a gallon of milk or milk by-products per day per hog, feeding a variety of grains instead of just corn, and making use of other feed sources such as vegetables and root crops. Barley, wheat, oats, and rye all have more protein than corn.

Minerals

Feed formulation is still relatively simple. It's a matter of finding the nutrient requirements of pigs and the nutrient composition of the feeds available, and doing a little arithmetic. It also includes such factors as nutrient interactions, cost, palatability, use of additives, variations in nutrient content of feeds in local areas during different growing seasons, and more. Then there is the matter of vitamins and minerals, which complicate the arithmetic even more.

Of the minerals that make up the body (and which therefore are of obvious importance for health and growth), calcium and phosphorus account for more than 70 percent. A deficiency of one or the other can result in poor gains, rickets, broken bones, or posterior paralysis. A large excess of either can interfere with the absorption of the other, so too much can have the same effect as too little. Swine need 1.2 and 1.5 times as much calcium as phosphorus. Since grains have more phosphorus than calcium and grains are a staple in the hog diets, swine are more commonly deficient in calcium than in phosphorus. This is further complicated by the fact that more than half of the phosphorus in grains is in the form of phytin phosphorus, a form that is poorly utilized by hogs.

Iron and copper are necessary for hemoglobin formation and for the prevention of nutritional anemia. Since sow's milk is very low in iron and copper, suckling pigs are especially prone to be deficient in these minerals. Once pigs begin to consume feeds other than milk they usually get enough iron and copper, especially if they have access to pasture.

Zinc is an important mineral for hogs because a zinc deficiency can cause parakeratosis, poor growth, and low feed efficiency. High levels of calcium and/or phosphorus in the diet appear to result in a higher zinc requirement.

Sodium and chlorine occur in the fluids and soft tissues of the body and play vital roles. Salt deficiency results in slower gains. Swine need less salt than do other classes of livestock. Usually ½ pound or less is mixed with 100 pounds of feed, but it can also be fed free-choice. By using a trace-mineralized salt, the feeder is also insuring against possible deficiencies in iron, copper, manganese, and iodine. These are not always supplied even by organically grown feeds raised on fertile soils.

Iodine is essential and may be deficient in the feeds grown on certain soils. Iodized salt takes care of the matter.

Commercially prepared rations are likely to contain ground limestone for calcium, and dicalcium, phosphate, deflourinated phosphate, steamed bone meal, or other low flourine phosphate material to supply readily available phosphorus. Homestead feeds such as milk and legumes or rape are rich in calcium. Tests have shown that the phosphorus from grains and other plant sources is adequate for satisfactory growth.

Other minerals needed by swine are found in sufficient amounts in natural foodstuffs. These include cobalt, magnesium, manganese, potassium, sulfur, and selenium.

Vitamins

The other major component of feeds is vitamins. Vitamin A occurs as carotene in plants and is converted by the animals. Yellow corn is a source, but an unreliable one. Lush pastures or green leafy hay are usual sources of vitamin A, and pigs on pasture are adequately supplied. Vitamin A is usually added to swine rations, and it's necessary for hogs fed grain and supplements.

Vitamin D is essential. Except for hogs raised in confinement, enough comes from sunshine. Vitamin D is needed for assimilation of calcium and phosphorus and is essential for normal calcification of growing bone. It is supplied in commercial rations by adding irradiated yeast, but if the hogs are allowed outside, the ultraviolet rays of the sun provide plenty of vitamin D even in winter.

The B vitamins are added by means of animal or marine proteins such as tankage and condensed fish solubles, and in distiller's dried solubles. Increasingly, the pure vitamins are added. Once again, the pasture scoffed at by modern, progressive hog farmers takes care of the situation in nature's way.

There is one exception Vitamin B_{12} is found in no feeds of plant origin except comfrey. (While alfalfa does not contain B_{12}, hogs on alfalfa or other good pasture are not deficient in this vitamin.) Vitamin B_{12} is especially important for young pigs.

The following B vitamins are also important:

Riboflavin(B_2). This water-soluble, B-complex vitamin functions in the body as a constituent of several enzyme systems. Therefore a deficiency results in a wide variety of symptoms including loss of appetite, stiffness, dermatitis, and eye problems. Pigs may be born dead or too weak to survive, or poor conception may result. Requirements of riboflavin appear to be higher at lower temperatures. A range of 1.0 to 1.5 mg per pound of ration is recommended, but cereal grains are poor sources.

Niacin. This plays an important part in metabolism as a constituent of two coenzymes. Loss of weight, diarrhea, and dermatitis are common deficiency symptoms. The niacin in cereal grains may not be available to swine because it is in "bound" form, but on the other hand the amino acid tryptophan can be converted to niacin. The National Research Council recommends 5 to 10 mg of niacin per pound of feed for growing pigs.

Pantothenic Acid. This B-complex vitamin is a constituent of coenzyme A, which functions in cellular oxidation of food materials. That sounds important, and it is, because it affects growth and health. The requirement is 5 to 6 mg per pound of feed.

B_{12}. B_{12} is essential for metabolism and grains and plant products, except comfrey, are poor sources. Only 0.01 mg per pound of feed is required, so it's not the sort of thing you shovel into the feed trough.

Other important vitamins are either present in normal rations or synthesized by the pigs: vitamins C, E, K, thiamine, pyridoxine (B_6), choline, and biotin. The last word isn't in on these either, though. For example, recent research indicates that additions of choline to sow rations may be beneficial.

We should discuss antibiotics and other additives, if only because they are so widely used in swine rations. They are not nutrients, but they have some of the same effects.

Antibiotics

Antibiotics, for instance, generally increase average daily weight gains, improve feed efficiency (meaning it takes less feed to produce a pound of pork), improve uniformity of performance (which is important to commercial producers), and may reduce death losses during the growing period. No one knows for sure how they do all this, although they apparently have an influence on intestinal bacteria. For starter ration, 40 grams per ton of feed is the usual dosage, 20 grams per ton are used from weaning to about 100 pounds, and 10 grams per ton after that. They are not beneficial to breeding animals. There are some indications that antibiotics improve the health of unthrifty pigs or that they control bacteria that compete with the animals for vitamins and other nutrients. Healthy pigs respond less to antibiotic feeding. The organic feeling is that animals fed antibiotics routinely have less natural resistance, and that humans consuming the resulting meat are also eating antibiotics and are therefore reducing their natural resistance. The question seems almost academic to the homestead hog raiser, because antibiotics are almost impossible to add to home-mixed rations and have no value under homestead management anyway. The same can be said of antimicrobial compounds, which inhibit specific harmful microorganisms in swine and improve feed and growth efficiency. Arsenical and nitrofuran compounds are widely used in commercial premixes.

Research is being done on hormones and enzymes. So far, these have not shown much practical value for swine so even the commercial producers don't use them.

These details probably don't have much meaning for homesteaders who don't have a laboratory in the basement and a couple of college degrees to go with it. The basic message is very important. All food is not the same. A pig cannot live on corn alone, much less thrive—even corn and pasture or barley and soybean oil meal won't do the job. Certain basic nutritional requirements must be met. Our forefathers did a pretty fair job through experience, intuition, and good sense, even without a great deal of technical knowledge. We know more than they did about certain technical aspects even if we aren't scientists or nutritional experts. Our problem as homesteaders is acquiring their experience, using our common sense, and combining that with what we know of the scientific aspect.

With that course, who knows? We might come out ahead of both grandpa and the purely scientific experts.

Then we come to the "unidentified factors," which are still controversial. As noted, these seem to be present in soil primarily, but also to some extent in forage (especially comfrey) and milk.

Hogs are well-known for their ability to balance their own diets, if given sufficient opportunity by providing a choice of feeds. Neither will they make hogs of themselves and overeat, unlike most other farm animals. These two factors, along with the principles of using feeds from several different origins, make homestead hog feeding relatively foolproof even without in-depth knowledge.

Remember also that growing and finishing pigs don't make very efficient use of forage, important as it is for nutrition. Grain is a necessity, although roots and vegetables can replace a portion of it.

A MODEL FEEDING PROGRAM

The young pig should get as much comfrey as it will eat and at least a gallon of skim milk, buttermilk, or whole milk per day. In addition, the pig should have access to ground grains, and preferably a selection of them such as corn, barley, wheat, and oats. You can figure that a pig will eat from 4 to 6 pounds of feed per 100 pounds live weight. Your weanling pig of 40 or 50 pounds will probably eat about 600 pounds of concentrate to reach a slaughter weight of about 220 pounds.

Give the pig kitchen wastes (not poultry bones or raw pork), cull, and excess garden and orchard produce.

As root crops, pumpkins, and other crops mature, these can replace more of the grain.

Some producers prefer to limit feed after the hog reaches 120 pounds. This increases feed efficiency and is said to improve carcass quality or the lean-to-fat ratio. Limited feeding can be estimated at 70 to 90 percent of full feed or the amount eaten in 20 to 30 minutes. This should come to about 1 pound of feed for each 30 pounds live weight.

Feeding swine organically by homestead methods is less efficient during the winter, and more difficult, especially if milk is in short supply. Milk is a major source of nutrition in the plan just outlined, and 1 gallon a day is a lot of milk for many homesteads, even during the summer. Comfrey and pasture in general are also very important in homestead hog raising. If these are available at all in cold climates, they will be in dry form and therefore of less value. There will not be the abundance of vegetables and other summer feeds.

If the pig is fed only grain, and particularly if the ration consists of a single grain, the homesteader had best follow commercial feeding practices by including a protein supplement. This will replace the protein and calcium of milk with soybean oil meal, tankage, and similar products. It will also include vitamins (which is all right) and antibiotics (which isn't all right).

Even the best alternatives are inefficient. Pasture can be replaced, in part at least, by well-cured legume hay and by cured comfrey. Both of these obviously entail more labor and/or more expense than the fresh product, with less value. Roots and pumpkins can be stored and fed to supplement grains, but they deteriorate in storage and storage involves double handling.

While it shouldn't make a difference in actual practice, there is less sunshine (vitamin D) in winter in many locations. Lurking behind all of these procedures is the fact that the pig will require a great deal of feed just to keep warm and the fact that you'll be battling frozen water and snowdrifts to do the chores.

Depending on the timing, you'll be faced with the job of butchering just at the time you should be getting into all the spring homestead chores. Some people like to raise their pig during the winter, but not us.

We could go into much more detail on the nutrition of hogs: the various nutrients they need at different life stages and the nutrition available in various feeds. Important as these factors are, they're much too complicated to be of practical use for homesteaders. Moreover, homestead feeding that includes milk or milk by-products, pasture, comfrey, a variety of grains—a variety of ingredients in general—is almost certain to be successful.

According to some sources, nutritional deficiencies among hogs are becoming increasingly common. A little introspection will provide some clues as to why.

One swine authority, writing in 1952, cited "leached and depleted soils" as one of the reasons for nutritional deficiencies, and he was not a spokesman for organic farming. The problem is even more serious today and will be considered especially so by those convinced of the efficacy of organic methods.

Therefore, one of the most important considerations in feeding livestock of any kind (and yourself too) is to use feed that was grown on fertile soil. Most homesteaders and small farmers are vitally interested in their soil, but building a healthy and productive soil on most of our abused farmlands is a long and arduous task. Pigs are of great importance in this concern because they are among the most prolific providers of barnyard manure.

Remember when we said that chemical farmers must use more chemicals to get a crop, while organic farmers are able to rely less and less on chemicals every year? The same endless circle is in evidence here. While hogs need feed grown on good soil, the feed grown on poor soil also produces less valuable manure. The rich get richer and the poor get poorer.

Chemically fertilized crops are not the same as those grown on naturally fertile soil, even though agribusiness experts like to point out that a plant can't tell the difference. This isn't the place to investigate that statement, but if you aren't convinced, consider for a moment that virtually all hog authorities believe the "unidentified vitamins and factors" come from the soil. There is too much we don't know to enable us to stray very far from nature's ways without a great deal of trepidation. At any rate, good feeds come from good soil, and deficient soils can only produce deficient feeds.

Forced production is another cause of nutritional deficiencies in swine. Very rapid growth, one of the major goals of modern commercial hog raisers, makes greater demands on bodies, and nutritional deficiencies of one element or another are more likely to occur. Breeding at early ages, farrowing at shorter intervals, and producing larger litters all require more nutrition than would be needed by animals that are not worked as hard. A part of this factor can show up on the plus side for homesteaders, however. Many nutritional deficiencies are minor and take generations to show up. There is greater leeway in nutrition in a hog you feed for four or five months and then butcher than there is in breeding stock that will be around a long time and must produce healthy individuals for future generations.

Perhaps the most important cause of nutritional deficiencies in swine today is confinement housing. The pigs have no sunshine, no access to real fresh air, and can eat only what the caretaker brings them. A meticulously formulated ration is of utmost importance in this situation, but even then there is considerable doubt whether the most knowledgeable expert knows everything there is to know. Just look at what we're still learning about human nutrition—and the controversies that surround much of it.

From reading this brief outline, it should be easy to see why the homestead hog raiser has so few problems with nutritional deficiencies. There are a few nutritional diseases that even the homesteader should be aware of.

Anemia. Many diseases affect baby pigs, which the one-hog homestead will not be involved with. One of the most common, for example, is nutritional anemia, usually caused by lack of iron. (It can also be due to lack of copper, cobalt, and certain vitamins.) It is most prevalent in suckling pigs that do not have access to soil. The symptoms are loss of appetite, labored breathing, a swollen condition about the head and shoulders, emaciation, and death. Hog farmers usually inject 150 mg of iron into the ham muscle at one to three days of age as a matter of routine. Most feeds contain adequate levels of iron for older pigs, but sow's milk is deficient in that element.

Hypoglycemia. Another disease affecting baby pigs is hypoglycemia, also known as baby pig shakes. Baby pigs with this are weak, shiver, and fail to nurse. Heart action is feeble and their hair becomes rough and erect. Death usually follows in 24 to 36 hours without treatment, which consists of providing warmth and force feeding 1 part corn syrup with 2 parts water at frequent intervals. It can be prevented by good management of the gestating sows, which includes adequate nutrition.

Osteomalacia. Lack of vitamin D, or inadequate calcium and phosphorus, or an incorrect *ratio* of calcium and phosphorus can cause osteomalacia, particularly during gestation and lactation. The ratio of calcium to phosphorus should be 1:1 to 1:5 for swine. Calcium is the most common deficiency in hogs because grains are deficient in calcium, and many pigs never get forages. The homestead pigs generally have no such problems. Symptoms of osteomalacia include stiffness of joints, failure to breed regularly, decreased milk production, lack of appetite, and an emaciated appearance.

Parakeratosis. On the other hand, too much calcium in the diet (above 0.8 percent) is no good either. It may result in parakeratosis, known by vomiting and diarrhea, reduced appetite, mangy appearance, and slower growth rate. This usually affects pigs from one to five months of age, which is the range for homestead hogs. Homestead pigs fed milk, legumes, and comfrey (which are all high in calcium) are prime candidates for parakeratosis. There might be some consolation in the realization that this is not a deficiency as such, but an overabundance, a poisoning. Furthermore, mortality isn't high. If it does occur under homestead feeding conditions, reduce the amount of forage feed and feed more grain. On a commercial basis add 0.4 pounds of zinc carbonate or 0.9 pounds of zinc sulfate heptahydrate per ton of feed. Parakeratosis isn't contagious, and the greatest loss comes from reduced gains and lower feed efficiency.

Salt Deficiency. Pigs need less salt than other classes of livestock, but they still need it. Salt deficient pigs may show depressed appetites, retarded growth, weight losses, and rough coats. They will also have a tremendous appetite for salt. Hogs may be fed salt free-choice, but if they have not had salt they will probably eat too much and encounter salt poisoning. That can result in death in a few hours or up to two days. The homestead treatment for salt poisoning is drinking large quantities of fresh water, although vets can administer water via a stomach tube or calcium gluconate intravenously if the animal is unable to drink. Beware of giving hogs access to brines from preserving and other homestead activities. One-half pound of salt per 100 pounds of feed is adequate, or if fed free-choice, take it easy at first if the pigs seem salt starved. Do not leave salt where it will get wet and form brine pools.

Selenium Poisoning. This is a problem in certain regions where feeds are grown in soils containing selenium. The pigs' hair falls out (and in severe cases the hooves slough off too), they go lame, and don't eat. Death is caused by starvation. The problem occurs in some parts of South Dakota, Montana, Wyoming, Nebraska, Kansas, and limited areas of the Rocky Mountains.

Blindness. Vitamin A deficiency results in night blindness and, in severe cases, blindness. Lush forage, green hay, yellow corn, and whole milk are all good sources of vitamin A.

Rickets. Rickets is caused by lack of vitamin D and also by a lack of calcium or phosphorus or an improper ratio of the two. It affects young pigs with bowed legs, enlarged knee and hock joints, and irregular bulges at the juncture of the ribs and the breastbone. Movement is painful, and swine frequently become paralyzed, but it's seldom fatal.

Iodine Deficiency. In the "goiter belt" of the Great Lakes region and in the Northwest, iodine deficiencies are possible because plants grown on soils in these regions are deficient in iodine. One symptom is pigs that are born hairless. Iodized salt is a simple preventative.

Fluorine Poisoning. This is another form of toxicity that can occur even on organic farms. The water in parts of Arkansas, California, South Carolina, and Texas contains excess fluorine and can result in such symptoms as abnormal teeth and bones, stiffness of joints, loss of appetite, decreased milk flow, diarrhea, and salt hunger. High fluorine phosphates used in some mineral mixtures may also be a cause. Fluorine is a cumulative poison, so its effects are more likely to be noticed in older animals.

Nitrite Poisoning. Nitrite, the reduced form of nitrate, falls into the same class. (This one is particularly interesting because of recent unfavorable publicity given to nitrites in bacon and other cured products. There seems to be poetic justice in the fact that the farm animals most susceptible to nitrite harm are swine.) In one study, four out of 10 farm wells in one Missouri area contained more than 10 parts per million nitrate—nitrogen, the maximum permissible amount for human water supplies. Some of them were as high as 190 ppm, which approaches the lethal amount for swine under certain circumstances. Excess nitrate (or nitrite) in water supplies is a common problem over much of the north central United States. Shallow wells hold more potential danger.

Hogs can tolerate nitrate but not nitrite; however, nitrite may be converted to nitrite. In shallow wells, coliform bacteria from surface contamination can convert nitrate to nitrite. The conversion can also take place in galvanized pails or other containers because of the presence of zinc.

The main problem appears to stem from nitrite interfering with proper nutrition even though the diet is adequate. This interference results from the action of the nitrite on vital enzymes and endocrine systems.

One researcher maintains that most nitrates in wells originate in organic matter on the surface: manure piles, straw and hay stacks, and other places naturally high in nitrogenous substances. This doesn't satisfactorily explain how almost half the wells surveyed in central Missouri in 1962 were hazardous according to government standards. What about nitrogen fertilizers, the use of which has been growing at a prodigious rate? At any rate, nitrate and nitrite are problems for some hog raisers.

In the main, nutritional deficiencies are problems for large-scale confinement hog raisers and test tube technicians. Homesteaders who use common sense, provide a variety of feeds, and good management needn't lose any sleep over their feeding program.

All controversies notwithstanding, there can be little or no disagreement that disease is simply the absence of health, and that health is maintained by proper diet, fresh air and sunshine, exercise, and a sanitary and comfortable environment. It might take some effort and knowledge to provide these, but not nearly as much as it takes to cure a sick pig.

In the improbable event that you encounter a sick hog, the wisest course of action is to call a vet or, at the very least, an experienced neighbor. Many swine diseases are difficult to diagnose without the knowledge and experience a one-hog homesteader might never accumulate or have to accumulate: you could raise one pig a year for 50 years and you'll only have experience with 50 pigs. A farmer who raised 10 or 100 times that many in a single year obviously will be and must be much more knowledgeable than the homesteader, but even the farmer doesn't engage in home doctoring. The farmer's job is to keep the animals healthy. If he fails, he moves over and lets the doctor practice his craft.

The importance of control and prevention cannot be overemphasized, especially for the homesteader. If we examine suggestions for health programs listed in various government publications, it's evident that many of the hazards to swine health simply don't exist on the homestead level.

FEEDER PIGS

Early in the summer, buy two or three feeder pigs (Fig. 3-2). Two or three hogs do better than just one because they will eat more feed in order to get it away from the others. The more you can get the pigs to eat, the faster they will grow. When raised individually, a

Fig. 3-2. Feeder pigs are excellent homestead animals (photo by Buhnne Tramutola).

pig will eat what it wants and then walk away.

When you buy feeder pigs, get females. Have them bred so they farrow as early in the spring as feasible. All the time the sow is carrying the little pigs, she is also growing. (Gestation period is about three months, three weeks, and three days.) Then, by selling six to eight-week-old feeders for $15 to $40, you will be able to have your own pork, as well as meet expenses. If you can't raise and sell six or more feeders from one sow, there's something wrong—either you are a poor feeder or you have a poor strain of swine.

When we want to keep a sow for more than one litter, we pick out the one that raised the most pigs. We like to keep a young pig each year. They have a tendency to get too big and clumsy when they get older and bigger. Hogs also lose their selling value after they get over 250 pounds (which is what a sow should weigh after having a litter).

Cutting Feed Costs

If you live near a grain elevator that has a cleaning system, you can save on feed costs by salvaging the cleanings. They will probably give them to you for cleaning them up. If the elevator does the cleaning, you can probably buy the cleanings at a discount.

Another way to cut costs is to buy standing corn at farm auctions. Before the sale, ask if you may pick a row of corn. If the answer is yes, pick a row; count the number of ears and then multiply by the number of rows. This will give you an estimate of how much corn there is and how much to bid on it.

If you buy a field of standing corn and don't have a tractor (or can't borrow one) to harvest it, cut a bundle of corn. (Our corn cutter is a modified hoe. Cut off part of the handle, leaving 16 to 18 inches and you have a great corn cutting knife.) Lay the bundles across a sawbuck until the stack is about 8 inches thick. Take a piece of rope and binder twine and tie it around the center. Then stack it for a day or two. After the corn is hauled home, you can pick the cobs, which is a good wintertime job when you can get in out of the weather. If you have a corn cutter, cutting box or old silo filler, you can cut the stalks and use them as feed or mulch. You can also harvest corn this way when the ground is too wet to use a tractor or truck.

Homemade Feed Ration

Another way to help the feed situation is to plant a lot of squash, pumpkin, and sunflower seeds and use them to make a homemade feed ration. If you don't have a feed cooker, get an old cast iron bathtub and put it on a couple of I-beams or angle irons to get it up off the ground, so as to be able to build a fire under it. Pour in three or four pails of water. Smash several pumpkins and squash and throw them in. Then get the fire going underneath the kettle or tub and cook until the pumpkins and squash are soft. While still hot, put two or three buckets of "slop" in a barrel (wooden, preferably) and add some chopped or ground grain. Keep adding squash and grain until the barrel is three-quarters full. Mix the contents of the barrel with a large stick and then cover it with an old piece of plywood. By the time the pumpkin and squash have cooled, the grain should be cooked. It doesn't hurt to throw in milk if you have some available.

Curing Black Teeth

Throw a piece of soft coal, about the size of a melon or grapefruit, in the yard to prevent problems with black (needle) teeth-baby teeth that never come out. We've seen little pigs stand by a trough full of feed and squeal their heads off because their mouths were so sore from black teeth they couldn't eat. The teeth can be broken off or cut with a pair of wire cutters. Pigs will break off black teeth themselves if given a lump of coal. There is something in the coal they need or like.

Getting Rid of Lice

Here is a good way to get rid of lice, which are a needless

expense. Get a piece of pipe about 1½ to 2 inches around, and about 4 or 5 feet long. Drill some small holes in the pipe, starting at one end and working down to about the center. Drive the pipe about halfway into the ground and wire some burlap around it. Fill the pipe with old crank case oil. The oil will seep into the burlap, and the pigs will rub up against it. Lice can't stand oil. Occasionally, fill the pipe again, so the burlap stays saturated.

Farrows During a Cold Spell

If your sow farrows during a cold spell, it pays to put the newborn in a small box and take them into the house (or someplace) to dry them off and warm them up. After they are dried and have a belly full of milk, they can stand a lot more cold.

Before the sow farrows, fasten a 2 by 12-inch plank around the walls of the farrowing house, about 12 inches to 16 inches off the floor. Then, if a sow lies down next to the wall while there is a little pig between her and the wall, the little one will have some protection against being crushed. (Poles can be used for the same purpose.)

Pigs are one of the cleanest domestic animals. If they are kept in a small, 2 by 4 foot pen, where after the first rain it is just a quagmire, they can't be clean. If you have them in a good-sized yard, they will generally pick a far corner in which to relieve themselves. A cow, horse, goat, or fowl will relieve themselves right where they are eating.

SUMMERTIME HOG RAISING

Summer is the time to enjoy the outdoors. Even the most pasty urbanite takes a pleasant stroll on a summer day. For hogs, though, summer is not the best of times. They are just not built for it.

This is easy to understand when one considers that a hog has a large layer of fat, a tough, thick skin, and a coarse coat of hair, all meant to hold in body heat. Anyone who has seen a hog nestle comfortably into a just-thawed puddle of March snow knows that hot summer weather is for girls in bikinis—not porkers.

A hog who is comfortable in summer is a hog who produces more. Feeders are more efficient if they are not hovering on the edge of prostration. Big sows are more patient nursers if they are not roasting in the sun. We have no definite proof, but hogs are enough like humans for us to suggest that boars are more active if they are cooler.

The ancestors of our present day domestic hogs were creatures of the Eurasian forest. Why not just let them run in the woods? First, not everybody has woods. Second, not everybody cares to run fencing through a tangled forest. Third, not everybody wants to ruin a woods. And that is just what a hog will do.

Ever wonder how the pioneers went about the monumental task of clearing the great northern forest for cultivation? They ran hogs on it. The porkers rooted up everything they could, leaving only the large trees that the settlers wanted to cut for buildings anyway. In the short term, there was no better place to raise hogs than the homestead woodlot. Anyone who has any consideration for woodlands should not let this violence occur.

Simple "Forest"

Why not duplicate forest conditions? This can be very simple. First of all, hogs can be kept on the northern edge of a woods. Since the summer sun does get high in the western sky, some protection in that direction helps during the late afternoon heat. Likewise, the shady side of any farm building is a good place for a hog pen.

If hogs are raised in a pasture without any trees or permanent buildings to provide shade, some sort of shade shelter should be erected. The simplest is the old reliable 4-H suggestion. Stick posts in the ground, frame a flat roof, then roof with boards. Top this with 2 feet of loosely piled straw to keep the heat off, and that's it. To make it portable, put a stringer across the bottom instead of burying the posts and drag it around like an old-time wooden sledge. A less expensive shelter can be built by using aluminum trailer skirting for a roof and logs for frames. Skirting can be purchased in 6 by 6-foot sheets and is lighter and more durable than plywood or planks. If you have to buy materials for a hog shelter, there is nothing cheaper. The framing has to be solid. Hogs love to rub against posts. It seems they get a kick out of toppling anything poorly built.

As far as size, shelters should be high enough so the tallest hog in the yard can't reach up and chew on the roof. Don't forget that feeder pigs grow taller as well as wider. A shelter 4 to 5 feet high will allow some air to circulate through the cooling shade and still be out of reach. Allow 7 to 8 square feet of shaded area for every hog.

The backyard hog raiser who keeps a couple of feeders in a small pen every summer can shade that pen by planting sunflowers on the south side. As the summer grows hotter, the plants grow taller and are more leafy, providing the greatest amount of shade just when it is most needed. Don't plant too close to the fence. Hogs love to chew on sunflower leaves. A double row 3 feet away from the fence will provide shade and protect the plants. At harvest time throw the stalks to the hogs. They'll be happy to shred them.

There is probably nothing more important to a hog in summer than plenty of fresh water. Anyone who keeps any animal is mistreating it if it doesn't have all the water it needs. A hog consumes about 1 gallon of water a day per hundredweight. Dependent upon the size of the litter, a nursing sow will drink up to 5 gallons a day. Don't worry about how much; just make sure they have it whenever they want it.

No one will ever accuse a hog of excess delicacy when it comes to how it puts things into its mouth. Therefore, the problem with watering hogs is how to keep the water clean. Metal troughs and buckets with open tops fail here. Tank type waterers, similar to gravity-operated chicken waterers, have the same drawback of the constantly muddy drinking trough.

The best method for watering a hog is with an open cup fountain. These can be attached right to the homestead water system and will provide clean cold water whenever the animal desires. If the hog yard is too far away from regular water lines, a gravity-feed fountain can be mounted on a tank. A careful handyman could install a fountain on a 55-gallon drum.

We use an antique galvanized steel hot water tank purchased for a dollar at an auction. Its side was already tapped, and the 1-inch connecting pipe from the fountain easily threaded through it. The tank is anchored solidly outside the fence, and the fountain protrudes into the hog yard. They drink at will by pushing on the flapper that opens the valve. It is a natural motion for a hog to push with its nose. We enjoy watching baby pigs learn to operate the fountain. It is one way of discovering which is the smartest pig in the litter.

Wallowing

Any discussion of hogs, summer and water must include the wallow controversy. Lounging in luxurious mud is the way hogs try to create heaven on earth. Coating themselves with mud is also the natural way hogs protect themselves from summer's insects. Though wallowing helps prevent blackflies from biting, it will not hinder lice or the mites that cause mange. A vigorous rubdown with used motor oil helps control lice, while spraying with Lindane is the accepted procedure for eliminating mites.

The biggest problem is that wallows are the breeding ground for internal parasites. Many swine worms spend part of their life cycle in the soil. A wallowing hog exposes itself to these parasites.

We have resolved the problem by letting them wallow after weaning. We butcher our own hogs, use the intestines for sausage, and eat the liver, kidney and heart. We examine the innards for worms. To our nonprofessional eyes, our hogs appear worm free. So, because it helps control external pests, we gamble with internal ones. But we hedge our bet in two ways. First, our sows always farrow on new ground. Youngsters aren't exposed to worm-egg infested soil that way. Second, we worm twice a year. Perhaps that is what animal husbandry is all about. Let the animal take care of itself as much as possible. Intervene only when absolutely necessary.

There is one final tip for successful summertime hog raising. On the hottest day of the year, steal a watermelon from the garden. Chop it up and pitch it over the hog fence so that everybody gets some. You may not believe it, and it probably doesn't matter, but you'll make a herdful of friends for life.

COLD WEATHER PLANNING FOR PIGS

Cold weather requires some planning for homestead hog raisers. In the area of proper nutrition, thought and extra effort must be given to available supplies of fresh ice-free water at all times for all classes of hogs, the baby pig, the weanling, growing-finishing, gestating, and the lactating sow.

The growing-finishing pig will roughly require 2 to 2.5 pounds of water for every pound of feed consumed. The gestating and lactating sow will require 4.5 to 6 gallons per day.

Watering equipment that will continue to flow during below zero weather is a must.

Daily supply of fresh platable feed sometimes becomes a problem during extremely cold conditions. Ice forms in feeder cups, feed will not feed from feeders properly, and pigs are automatically placed on a limited ration. Consequently, gain is restricted.

Another problem is the feeder allowing too much feed. Pigs become wasteful and the feed is pushed out on the ground. Chances for consumption are small. Proper adjustment and feeder check at least three times weekly will generally solve these problems.

Every 10 pounds of feed wasted each day at the present 6¢ per pound feed price amounts to $219 per year. Ten pounds of feed is often wasted each day with a poorly adjusted feeder. Worn-out feeders and overcrowding of feeders also add to feed loss.

Farrowing units must be warm, comfortable, and dry. A farrowing house temperature of 70-75 degrees Fahrenheit and a nest of

Fig. 3-3. Make sure pigs are fenced in properly.

sleeping area temperature of 90-95 degrees Fahrenheit for newborn pigs. Newly weaned and growing-finishing pigs must be made comfortable. A chilled pig is a high-cost pig. Wiring, equipment, outlets, and fuel supply should all be checked, made operative, safe, and ready to go during the sudden extreme cold period (Fig. 3-3). Throwing the switch on or firing the equipment when it has not been checked thoroughly since late last winter might be just the right steps to burn down a good facility and waste a barn full of live pigs.

Bedding needs and availability should be considered. The pig in a cold gravity ventilated building will do an outstanding job of converting feed and daily gain if comfortable. If not, the producer will be disappointed in performance but it's not the pig's fault.

What about the sudden three-day snow storm with 10- or 12-foot snow drifts? A planned program on how to operate at this time is essential. Feed supply, water needs, bedding, equipment to remove snow, and extra labor are all part of the top swine producer management program.

External parasites are sometimes neglected during the extreme cold weather months just at the time when these parasites are probably most active. Lice and mange control measures should be planned and worked into the schedule of events.

Movement of breeding and market hogs during the cold weather requires a little extra protection from the cold wind. Bedding that is the same as a blanket for man will help in pig comfort and will enable the pig to perform its mission. Avoid loading hogs up an ice-covered trail into a slippery loading chute; this will be a helpful management practice. Pigs that can't walk when they reach your home or market are of little value.

Looking for the "weak link" in your wintertime hog management phase of operation and correcting this weakness now will go a long way toward a successful wintertime operation.

PLOWING A GARDEN WITH PIGS

Hogs have this natural ability to bulldoze earth with their snouts as they search for goodies to eat. They can till up your garden plot, or turn a soddy spot into a brand new garden, or clear a weed-infested, even brushy, piece of land for use as pasture.

The garden, right after you've harvested all the vegetables, is an excellent place for the hogs to do their rooting. Not only will they work the soil, but they'll also add fertilizer in the form of rich manure. They'll also turn any cull produce they find into pork. We're all aware that former pig lots can be top-notch gardens. You can get the same results in your regular garden by a late fall visit from the hogs.

Fencing around your garden is a must. If there's already a fence around it, you must make sure it's really hog-tight. If you have to make one, woven wire hog fencing, 32 or 36 inches high, is one way to go. (The woven wire will also keep out garden pests during the growing season and even act as a support for your vine plants.) If you're the owner of an electric charger, an electric border will do the job, and it's the quickest way.

You may also have to put a length of barbed wire or nail boards along the bottom of the fence, especially if your garden is on the small side. The smaller the plot, the more the pigs want to get out. The little garden will also need a higher fence than the larger plot.

If you've had hogs escape from their pen, you know why a good fence is needed. Chasing them can be like chasing wild foxes, and some individuals even seem to enjoy giving you a run. An occasional loose one isn't bad, but when damage is done it's a different story.

You may also want to provide shelter for your plowers during their stay in the garden. The weather at the end of the growing season can be mean. Your animals would appreciate a warm, dry, temporary house to snuggle in, whether it be a portable plywood shelter or just a straw bale hut.

With fence and shelter up and your vegetables all picked, it's pig moving time. A pigpen that's close to the garden makes for convenience. You can put the pigs in a hog crate if you're not real acquainted with each other. You can also just open the pigpen gate

Fig. 3-4. These pigs have just been turned loose on a future garden site.

and lure them along with a pan of feed or, better yet, milk. It also helps if the pigs are a little hungry. If you just have one or two hogs, you'll have to use just them. If you have a choice, use several younger hogs or two or three larger animals to get the job done quickly. The smallest pigs you can use practically will weigh about 75 to 100 pounds or so, the size depending on the lightness of your soil.

Once the pigs are in their new home, they'll run around for a while and look things over. After that, they'll start chewing on vegetables and weeds and soon turn the soil (Fig. 3-4).

The more pig power you have on your plot, the quicker you can take them off. The right timing is rather important. You'll know it's time when most or all of the vegetation is eaten or buried and when the soil is still on the loose side (Fig. 3-5). Leaving the pigs on over winter is undesirable since the soil will get too compacted and the pigs will be adding manure too close to spring planting. The fertilizer must be decomposed some for the well-being of your new vegetables.

Your hogs can also be used to plow up sodded ground for use as a new garden or small pasture. For this project, you can use larger hogs only. It takes muscle to turn over that thick-rooted sod.

The practice of pig plowing is as good for the homesteader as it is for the pasture or garden. You save time and labor, and you're also saving money by not buying the gas you would have used in your tiller or tractor. You may not have to use these machines at all.

At the same time, while you're providing the pigs with their usual amount of water, you won't need to give them as much feed if

Fig. 3-5. This site has been plowed and fertilized by pig power.

there are plenty of plants for them. If the pigs were going to get the waste vegetables anyway, letting the animals help themselves means you don't have to pick or haul. You'll also have fewer weeds to be bothered with come spring.

The hogs are way ahead, too. They get to eat plenty of vegetables and other plants and take in valuable minerals from the soil. They also get fresh air, exercise and a naturally sanitary living space, and the chance to root to their heart's content.

ARTIFICIAL INSEMINATION

Do Sunday morning fantasies of sizzling homegrown bacon have you contemplating introducing a hog on your homestead? If so, you might want to consider the possibility of keeping a sow rather than fattening an annual round of feeder pigs. In addition to providing a year round supply of pork, plus a few feeder pigs to sell instead of buy, the old girl can plow and cultivate the garden as well as recycle garden wastes in the form of manure. The money we've made from the sale of piglets pays for all the feed the pigs eat. Consequently, although we raise no corn or soybeans, we eat home-raised pork.

The biggest problem with this arrangement, though, is getting, the sow bred. The experts advocate maintaining a closed herd to reduce the danger of transmitting disease or parasites. A one-sow farmer can hardly consider maintaining a boar. Even if you can find a neighbor who is willing to lend you his boar, the animal may not be the right size (a VW cannot mount a Cadillac) or may be of poor quality.

The answer for us was artificial insemination. Artificial insemination eliminates the problems of finding a boar as well as the stress that is incurred when transporting the animal. It also adheres to the experts' advice of maintaining a closed herd. Last, it allows you to choose semen from champion boars that you would otherwise not have access to. Upgrading stock is one of the most valuable objectives any herdsman can have.

Swine insemination does not require special training. What it does require is diligence. A boar and sow running together know the proper time to breed. When the boar is replaced by artificial insemination, the breeder has to determine the correct breeding time.

Detecting when a sow comes into heat depends on careful observation of the animal's estrous cycle. Ironically, this most important factor in successful breeding is also the most variable.

Heat Cycle

A hog comes into heat on the average of every 21 days—so they say. The heat cycle actually ranges from 18 to 24 days. Before attempting to breed an animal, her particular cycle should be determined as accurately as possible.

A hog's heat cycle is divided into three phases: *proestrus, estrus,* and *diestrus.* Proestrus and estrus last about two days each and are of great importance to the breeder. During proestrus a sow becomes restless, paces the fence, and frequently urinates. Gradually her vulva turns pink and begins to swell. The signs differ somewhat for each animal. We once had a Duroc gilt whose swollen vulva resembled an orange tennis ball complete with fuzz. Other animals will doubtlessly have less dramatic characteristics. Another noticeable sign is a gradual lengthening of the slit of the vulva as estrus approaches.

Examine a sow at least once a day for two months prior to breeding. This is not very complicated. Simply approach the animal to be bred while she is eating, go around to the back end, lift her tail, and look. A calendar recording daily observations is a great help.

It is only during estrus, or standing heat, that a sow may be bred. At this time her vulva will be swollen to its greatest extent, and she will have a clear vaginal discharge. This discharge is the clearest sign of the correct time to breed. She will also stand solidly and let the herdsman straddle her back. Her ears will cock forwards, as if she were attempting to hear a faraway sound. Ovulation occurs 36 to 40 hours after standing heat begins. The hog should be inseminated 14 to 22 hours after the onset of estrus. When semen is introduced into the hog at this time, the greatest number of sperm contact the greatest number of ova, resulting in the largest litters.

Boar semen comes in two forms—liquid and frozen. The liquid has a shelf life of 60 hours and must be stored until used at 42 degrees Fahrenheit. Frozen semen is pelleted and kept in a liquid nitrogen tank until breeding time. The pellets are then thawed in a precise manner. A reusable kit, consisting of small Styrofoam boxes, a beaker, and a thermometer, can be purchased for $12. This helps simplify the operation. Thawing takes about 10 minutes and, if you follow directions carefully, is actually quite simple. The semen is then ready to use in liquid form.

Equipment

Insemination equipment is neither elaborate nor complicated.

It consists of a rod, adapter cap, and bottle. The plastic rod is 17 inches long and ¼ inch in diameter, with a bent tip. The straight end inserts into the 2-inch rubber hose of the adapter cap. This cap screws onto the plastic bottle containing the semen. The rod and bottle are disposable. The price of the rod is 4¢, and the bottle is free with the purchase of semen. The adapter cap cost 20¢. It should be thoroughly rinsed and stored in a clean, dry place after each insemination.

The actual introduction of semen into a sow appears hazardous, but it is not (Fig. 3-6). The pig should be in a small pen by herself. She should not be tied. One useful trick is to hold back her feed several hours. Making breeding time feeding time blissfully distracts her from the activity at her stern.

The semen bottle with adapter is fitted onto one end of the insemination rod. With the bottle end held below the level of the vulva, the curved tip of the rod is slowly inserted. The rod carefully shoved farther in until the obstruction is felt. A slight twist or slow wiggle will push the rod past this slight block, the entrance of the cervix. After penetrating farther (about 12 to 14 inches of the insemination rod should be in the pig), squeeze the semen bottle. Sometimes there is an air pocket that holds the semen in the bottle, making it very difficult to squeeze. Moving the rod slightly usually enables the semen to flow properly. Squeeze the bottle until just before it is empty. Leave some semen in the rod and bottle to eliminate blowing air into the hog and forcing semen out. Some semen may drip out the vulva anyway. We have never known this to do any harm. Removal of the rod completes the operation.

Breeding should be repeated in 16 to 28 hours. Provided the sow is ready, the whole insemination can be completed in 10 minutes. If the sow isn't in standing heat, don't even try.

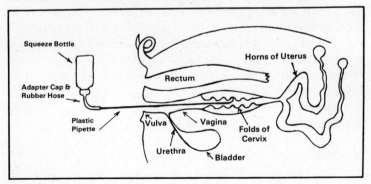

Fig. 3-6. Introducing semen into a sow.

We learned the artificial insemination procedure by reading the directions provided by the breeding association. No experience is necessary for any literate person to be able to successfully breed a hog.

To obtain semen, contact a breeding co-op listed in the Yellow Pages. County agricultural agents can certainly be helpful.

HELPING A SOW GIVE BIRTH

When your sow is ready to farrow, are you prepared? Have you taken the proper steps to help her in her hour of confinement? No doubt you've read all the material available on swine and sow care, but if this is your first experience, you are likely to be nervous.

Extra attention during and immediately after the birth process will pay off in dollars. In red ink on the calendar, circle the sow's farrowing date, which will be three months, three weeks, and three days after she's been bred. (You can also figure the date by consulting any gestation chart.) A week before the actual farrowing day, shut the sow in her freshly strawed pen. It's important that she be relaxed and adjusted to her place of confinement ahead of time. Otherwise, she may decide to make her nest in the woods or under the front porch, especially if she's used to running loose. Sometimes a sow will decide to deliver her babies ahead of your schedule. If she's safely incarcerated in her pen, you can avoid a potential disaster.

Prepare for the event by stocking a box or basket with essential medical items. In the box put a bottle of iodine and a small jar (a baby food jar works well) to pour the iodine into. Include a short-bladed, large-handled pair of scissors to cut navel cords with. A bunch of soft clean rags are a must. Old terry cloth toweling is especially nice. Enclose a vaccine syringe and a bottle of Combiotic in a plastic sack or inside a fruit jar to keep them clean. Also include in your kit extra and clean cotton work gloves because the ones you're wearing may get wet and slippery. Lastly, but very important, put in a roll of toilet tissue.

The entire birth process can take from two to as long as 24 hours. There are times when it will be necessary to wait all night for the final countdown. If the temperature is warm, the only problem for you is boredom. This can be alleviated by taking along a selection of reading materials. If it is very cold and you have to wait, try wrapping up in an electric blanket. This can be hard on blankets, but if you save a couple of baby pigs by being present, you can buy a new blanket from the profits.

In due course, the sow will grunt, moan, contract extra hard, and out will swish a piglet. It may come head first or in breech presentation. There seems to be no set rule. Use one of the soft rags and towel the little one dry. Scissor off its dangling navel cord leaving only an inch or so.

Using the small jar filled with iodine, douse the cord into the iodine. Then park the piglet's nose in front of a faucet. Sometimes the piglet's navel cord is already short, but more often it is long and can become entangled. The sow, in changing positions, can step on the cord, immobilizing her offspring. When the little one can't get away, it is apt to be fatally smashed by the sow.

Ordinarily, a fresh born baby pig has only a thin membrane covering it and will immediately blink, gasp and begin staggering toward Mama's lunch counter. Sometimes, however, the individual piglet will be born enclosed in a thick membrane. Quickly tear off the membrane or the baby will suffocate.

Usually the sow will steadily push out her babies at regular intervals. Sometimes, though, she will reach an impasse, a hangup somewhere inside that slows the pace of delivery. She seems instinctively to know what to do. Generally she will lurch to her feet, turn around, and lie down on her other side. The action seems to promote fresh contractions. It is during those times when the sow struggles to her feet to shift position that the baby piglets court suicide by rushing about under her feet and belly.

If the sow is gentle, scoop up the babies and park them under the heat lamp in the creep. If she is grouchy, use an instrument like a padded sheephook or rake to whisk the babies out of harm's way. If you're scared to crawl in with her, and it takes her a long time to settle back down, keep the babies shut in the creep. Use a board or have a flap door that can be lowered attached to the creep opening. When the sow settles down and seems content, let the little ones out. The more they suck, the more colostrum they'll take in. As with all newborn the mother's milk acts to immunize the baby against diseases.

When at last the sow begins to discharge her afterbirth, you can begin to feel proud. Generally, some afterbirth material will be discharged. One or two and perhaps more piglets will be born. Watch for these. Sometimes they are pushed out encased in a heavy membrane. When you are sure the last baby has been born, administer 10 cc's of Combiotic. If you insert the syring needle in the fatty tissue between her back legs, she won't realize what you're doing—and won't make a fuss.

Pick up your tools, equipment, and soiled rags and leave. Neither the sow nor any of her 14 kids are paying the slightest attention to you.

Feed grain very lightly in the first 12 hours after the sow delivers, but give her plenty of hay. It doesn't hurt to give her approximately a cup of mineral oil the first feeding after birth to keep her bowels loose. Sprinkle the oil on some grain or put it in a pan of milk.

SELECTING REPLACEMENT GILTS

Swine producers need to be careful when selecting replacement gilts, advises Ken Drewry, Purdue University extension swine specialist.

Sows in the swine herd primarily determine litter size at weaning. They also contribute half of the genetic potential of the pigs for growth, feed conversion, and carcass merit, he points out. Number of pigs weaned per sow and the ability of the pigs to grow and convert nutrients into quality pork economically are two important factors affecting economic returns for a swine herd, he adds.

Select the fastest growing, leanest, sound gilts from large litters as replacements. Pay attention to factors in the areas of sow productivity, gilt soundness, and gilt performance.

According to Drewry, sow productivity includes such factors as litter size in which a gilt is born, the ability of pigs in the litter to survive to weaning, and the weaning of the litter. Genetically, these factors are traits of the gilt's mother, not the gilt, he points out. These economically important traits do respond to selection, however, although the response is lower than with other production traits. This is because the management and husbandry practices of producers have a large influence on these factors, and because the selection is based on the record of the mother, not the gilt.

Gilt soundness includes such factors as normal reproductive development, number of functional teats, and skeletal soundness. All of these respond to selection, and the selection is on the basis of the gilt's own record. Drewry suggests that only gilts with a normally developed vulva be chosen, because a small vulva is usually associated with small or infantile reproductive organs. The replacement gilt should have at least six, evenly spaced, functional teats on each side of her midline, Drewry says. A gilt should not be chosen if she has feet and leg soundness problems that will interfere with breeding or her ability to farrow and nurse a litter of pigs.

Several performance traits may be measured on each potential replacement gilt; however, Drewry recommends that only weight and backfat be considered. These traits are easy and relatively cheap to measure, he points out. Larger genetic response can be realized for these traits, he said, because management and husbandry practices have a smaller influence, and selection is based on the gilt's own record.

When comparing weight and backfat measurements, weight should be adjusted for age. Backfat should be adjusted for weight.

The following gilt selection calendar is recommended by the specialist:

- At birth, identify gilts from large litters from good sows and which have 12 or more evenly spaced teats.
- At weaning, place gilts on good growing and developer ration; screen gilts identified at birth on underline, weight, soundness and conformation; and select two or three times the number needed as replacements.
- At five months, weigh and probe all potential replacement gilts and evaluate for soundness. Select the heaviest, leanest gilts as replacements, saving about 25-30 percent more than needed for breeding. Place selected gilts on restricted feed and give fenceline contact with boars.
- At breeding time, make the final cull, keeping sufficient gilts to offset the 15 to 20 percent infertility observed with virgin gilts.

THE SCIENCE OF CASTRATING PIGS

The male pig, with the exception of the individual animal being kept for breeding, is one animal that must be castrated before eating. A few people have eaten boars, but the meat is usually foul smelling while being cooked and has a taste that is about as bad.

It is much easier to castrate young pigs. Not only are they much easier to hold, but the blood vessels are much smaller, and bleeding is next to nothing.

Castrating

Baby pigs can be castrated any time after they are a week old. In general, it is best to wait until the pigs are well-started, active, and eating well.

There are special racks built for one-person pig castrations. Most homestead surgery is easiest and more cheaply done by calling in a helper.

The baby boars should be herded out of the sow's pen. It is easier to entice them out with a little milk and mash, then block the exit, than to catch them, one by one. Mama sows can get *very angry* at such human "attacks" on their babies. Often the first squeals cause a nice, quiet, pet sow to leap up to the defense, looking like a savage tiger. Don't think for a minute that the sow won't or can't use those teeth. She can chomp an arm off when riled up. Be careful.

Once the pigs are all away from the sow, sort out the gilts (young females) from the boars. This will make catching the pigs much easier, as you won't have to sort while surgery is going on.

With the boars alone in a corner, gather your equipment. For castrating baby pigs, you should have:

- One sharp scapel blade (a handle is not even necessary). You can get this form from your veterinarian at a very low cost.
- A pan of warm, soapy water.
- A bottle of scarlet oil, iodine, or other disinfectant. (Don't use Lysol or any other harsh disinfectant.)
- A strong needle and strong carpet thread. This is just in case you discover a ruptured pig.

Now you are ready to begin surgery. Have courage, as it is easy and it is not bloody and horrible. Have your helper grab a pig by the hind feet, holding it upside down, with its belly toward you. The helper can grasp the pig between his knees to steady it. The pig should not touch the ground with its front feet.

Using the soapy water, scrub the area just in front of the scrotum. Dry the area. With your left hand (reverse if you are left-handed) push the testicles forward as far as they will comfortably go. This is where the incision is made, right over one testicle, still held bulging against the skin. The incision should run lengthwise (head to tail) and be about an inch long. Squeeze the testicle quite hard, and it will pop out of the incision.

If it will not come out, deepen the incision a little, and it will. Don't cut it off. Instead, firmly grasp the testicle and cord and firmly pull until it comes free. The cord will be quite long. Using this method, there is very little chance of bleeding.

Push the other testicle against the incision, and gently cut against it, using the same skin incision used for the first testicle. Again, squeeze it, and it will pop out and be ready to pull free.

If there is any ragged tissue hanging from the incision, either pull it free or trim it off. Do not leave anything hanging out, or the pig may end up with a mass of scar tissue and infection.

Splash the incision with disinfectant and release the pig. With practice, this process is very easy. Watch for ruptured pigs.

The tendency to rupture into the scrotum is quite common in pigs. It is a hereditary condition, so it is wise not to knowingly buy a gilt, sow, or boar from a farm that has this problem. In rupturing, the ring through which the testicles descend from the body fails to close. This failure allows the bowels to protrude into the scrotum, along with the testicles. The pig may be only slightly ruptured or severely so. He may also only be ruptured on one side or be ruptured on both sides. Severely ruptured pigs can be distinguished before castrating.

In a normal pig, the testicles are definite and hard in the scrotum. The ruptured pig, however, has a larger scrotum, filled with softer material (the bowels), as well as the testicles. If you have a pig you suspect is ruptured, it is best to take him to a veterinarian for castration. He can castrate the pig and repair the rupture so that when the pig is sold, he will not look like a boar, bringing a lower price. (In home-repaired pigs, the pig will live and grow and be fine to eat, but will have a scrotum full of bowels, looking like a boar.)

In many cases, the pig will only rupture after the testicles are removed. This is because they held the ring closed partially, preventing the bowels from entering the scrotum. These pigs are the tricky ones. Unless you look at each one after castration, you could miss a ruptured pig, turning him back into the pen. In such a case, he will soon be dragging his bowels around in the pen. Even such a pig can often be saved by rinsing the bowels off in warm water, carefully stuffing them back through the correct ring, then suturing the scrotum up. It is much safer and better for the pigs to check each one before placing it in the pen.

In the normal pig, you can look through the incision and see a vacant space, with only smooth, pinkish rounded muscle on the inside of the thighs and belly. The ruptured pig however, has torturous, dark, almost purple, bowels, which are about as big around as a lead pencil, visible on one, or both sides, near where the leg joins the body.

When a ruptured pig is discovered, it is best to have a veterinarian make the repair. This is seldom possible, though. If left open, the bowels will soon slide out of the body, and the pig may go into shock. When the veterinarian is quickly available, keep the pig upside down and get to him immediately. If not, you can close the incision after castration, saving the pig.

A curved, carpet, leather, or surgical needle is best, but a darning needle also can be used. The needle should be threaded with a strand of heavy thread, such as upholstery or leather thread. Lighter thread will cut into the skin and finally break or cause an infection.

Care must be taken not to hook a bowel when sewing the incision closed, for to do so is to kill the pig. The best stitch for such a closure is the mattress suture. Here you will pass the needle through the skin on first one side, then the other of the incision, then down ⅛ to ¼ inch, then back through the skin on the same side, then across the incision and through the skin on the other side. The two ends are then tied securely, but not overly tightly.

Usually, two such stitches completely close the incision. Don't overclose or neglect putting enough stitches in the incision. There shouldn't be a hole for the bowels to work through, but the oversutured incision will often fester and become very painful.

If the bowels have not been in contact with straw, the ground, or manure, antibiotics are usually unnecessary.

Castrating the Older Boar (100-200 Pounds)

Many people think that if a pig has been left a boar until he is larger, that he must remain so, and that he will be unfit to eat, even if he is castrated at this late date. This is not true. We have bought many young boars either for breeding one sow or just because they were cheap, used them if desired, then castrated them. After castration, they should be fed out for at least four weeks. These stags make good eating. In fact, we can't tell their pork from the meat of a barrow, castrated at two weeks of age. The pig castrated at a young age will, however, put on faster, and more economical gains, and be much easier to handle.

Needed to castrate the larger boar are:

- A scalpel (blade and handle), or very sharp jackknife.
- A length of #2 chromic catgut (available from your veterinarian) or an emasculator, which crushes the blood vessels, while at the same time, cutting.
- A pan of warm soapy water.
- A strong rope, at least 14 feet long.
- Two or more strong helpers.
- A bottle of disinfectant (not Lysol or bleach).

You will also need a strong, overhead beam available, or a tractor with a front-end loader.

By this age, you should be able to spot a ruptured boar. The scrotum is usually quite enlarged, and the testicles are not defined, as in the normal boar. If in doubt, though, either have your veterinarian castrate the boar, or have on hand the needle and thread.

The hardest part of castrating the larger boar is the restraint. Even this can be made much easier with a few hints.

First, have the beam or tractor standing by. Then quietly get in the pen while the boar is eating, and sneak the rope around one hind leg, just above the hock. Use a lasso-type slip knot, and again, quietly, draw the noose snug. Either tie the rope around the bucket on the tractor or throw the other end over the beam to your helpers. Then get out of the way. The boar should be lifted up, until his front feet are clear of the ground. When he is in this position, the helpers can tie the rope (being certain that it can be easily untied), and come help you.

One helper can grab each hind leg. Sometimes a twine on the leg will help. The area in front of the scrotum is scrubbed with soap and water, then dried. The same castration procedure is used with larger boars that is used in castrating baby pigs, but instead of just pulling the testicles free, you will have to tie each one off with the catgut to prevent bleeding. Catgut is necessary, as thread, cord or string will not dissolve, but will irritate and cause trouble later.

Be sure to enclose the blood vessels and cord in the catgut, when doing each testicle. Pull the catgut as tightly as possible. Making more than one wrap on the knots will make them hold better. Use at least two knots on each tie to guard against slipping. After the cord to the testicles has been tied tightly, cut the cord an inch closer to the testicle. There should be no bleeding. If there is, you didn't get the knots tight enough. A few drops of blood will not hurt anything, but steady bleeding should not be allowed. Retie the catgut, closer to the body, before removing the old tie.

Repeat with the second testicle. Drench the area thoroughly with disinfectant, and you are done.

Take care in releasing the boar—now a stag barrow. Don't get bitten or break his leg as he is lowered to the ground.

Suturing the scrotum is not necessary or advisable on normal boars. This only prolongs healing time and can cause infection. The same is true of small incisions. Many people think that they are doing the boar a favor by making a tiny incision. It does look nicer, but often quickly becomes infected. Tetanus is common with such incisions. Better to make a reasonably big cut, which will heal from the inside out and will drain, if necessary.

Incisions in castrated baby pigs will heal in about three days. Remove any stitches in ruptured pigs two weeks following surgery, sooner if healing is complete. The incision in a bigger boar will heal in about two weeks.

Boars larger than 200 pounds really should be castrated by a veterinarian, as there is more chance of severe bleeding. There is no size limit to castrating boars.

HOG SKINNING

The time-honored hog butchering and scalding bees in which neighbors joined together to lay in their winter pork supplies is not as appropo today as in years gone by, because most small family operations are not geared to this approach. In yesteryears, every farm had scaffolds, pulleys, scalding vats, pitchwood, scraping tools, and pig hooks. You spent a lot of time chopping wood and several hours heating water. You had to have extra hot and cold water available because the temperature had to be pretty close to exact in order for the hair to scrape. The correct temperature was usually determined by the hand test. If you could put your hand in the heating vat three times in rapid succession without scalding yourself to death, it was ready for the pig to be slithered in for scalding. It took a lot of back power to slide the hog in and out of the water and to use a hair pulling procedure. Sometimes a handful of lye was added to help slip the hair and to give the skin a cleaner, smoother look. Then came the scraping, which took quite a while because after the hair cooled, it would sometimes set and would be difficult to remove at best.

There are several ways to kill a hog. You can shoot them and then bleed them, stun them with a hammer and bleed them, or just use a knife directly and bleed them.

Sticking a pig for bleeding is a little different than bleeding most other animals. You make an incision with a long sharp butcher knife perpendicular to the neck and not across it. You make the stick in the throat just ahead of the shoulders and not up in the jowl or thorax area. The knife is inserted and then turned to sever the jugular. Have the pig's head downhill so rapid bleeding occurs. If it is uphill, the blood will generally filter into the meat. Put the pig on a piece of plywood, get a brush with some hot soapy water, and give it a scrubbing from top to bottom. Hose or rinse the soap away and you're ready to hang him up for skinning.

Cut beneath the tendons on the back hocks so you have a place to secure a gambrel. A gambrel can be a piece of wood or tree limb 3 feet long that is pointed so that the back legs of the animal can be spread apart for scaffolding and skinning. Hang a 2 by 6-inch board, or a heavy tree branch, between two trees, about 8 feet in the air and attach a pulley and rope. Tie the rope to your car bumper, attach the other end to the gambrel, and raise the pig into the air so its head is just off the ground (Fig. 3-7).

Now you need a retractable linoleum knife. This knife blade has three settings. Use the first setting because you want to cut through the skin and into very little of the fat. Cut from inside the hind leg across the crotch to the inside of the other hind leg. Now you are ready for the stripping. Start with the knife and cut strips 2 to 3 inches wide from the hams to the head, cutting only through the rind. Do the cuts completely around the body and always from the

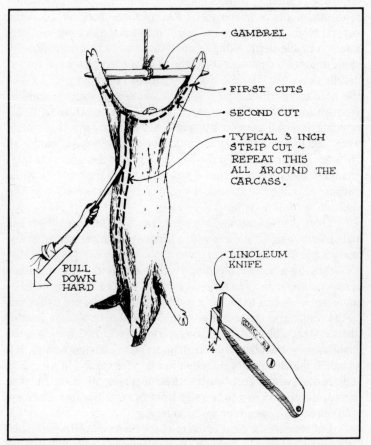

Fig. 3-7. Preparing a pig for butchering.

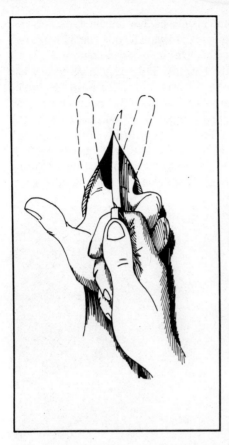

Fig. 3-8. Position of the hands when opening the belly.

top. Skin each strip a few inches and then pull by hand (Fig. 3-8). It will come off slick as a whistle, and you lose no meat and very little fat. When you purchase pork, the only pieces of meat you generally find with rind on are the bacons and hams. Hams and bacons smoke and cure as readily without the rind and are just as tasty.

After it is skinned, take a hatchet or cleaver and split the bones right in the center between the hind legs. This will expose the intestines and the anus. Cut completely around the anus and do not sever the intestine attached (Fig. 3-9). Tie a string around the end of it to prevent manure contamination of the meat. Now take your knife and cut gently down the stomach 6 inches. Then slit the skin from the brisket up to meet the other cut. After you have an opening put your index and middle fingers, palm up, in the incision. With knife in right hand, slide it upward, using these two fingers to hold the intestines away from the knife as it moves upward (Fig. 3-10).

Have a container to catch the insides as they fall out of the carcass. You will have to use the knife to cut away the interlining, lungs and heart in the throat region. Squeeze the blood clots from the heart and sever the liver from the entrails. Then cut the green bile sac from the liver as it will make the meat bitter. Take the hatchet or cleaver and using a hammer as a driver, split the thorax down to the jaw. Sever the head by cutting completely around and twisting the head sharply while another holds the body firmly. You can salvage the tongue and brains from the head. The head can be skinned, cooked, and used to make head cheese, which is a rich meat delicacy that makes very good sandwich meat. Now you can sever the spinal

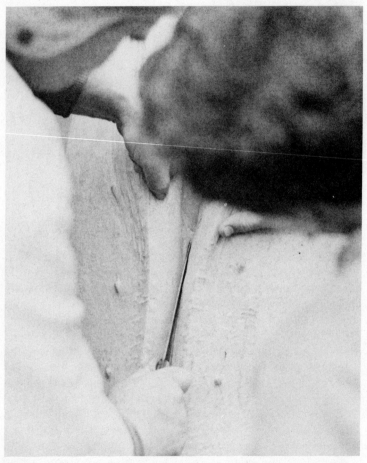

Fig. 3-9. The first cut is made along the belly. Take care not to cut the intestines (photo by Jean Martin).

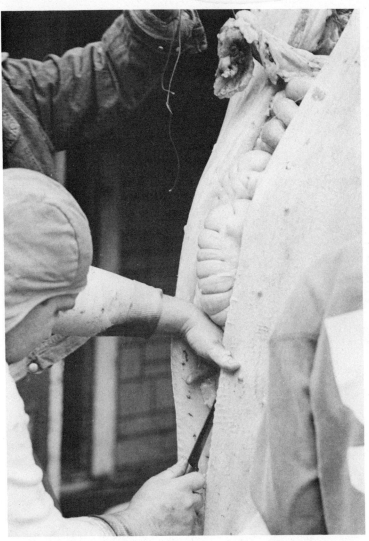

Fig. 3-10. As the cut progresses, holding one hand inside will help avoid cutting into the guts (photo by Jean Martin).

column in half using the cleaver and the hammer (Fig. 3-11). The butchering is done. The cooling process generally takes 24 hours.

BUTCHERING PIGS

This method is very simple and fast. Two experienced men can do the job in an hour. The principle steps are as follows:

- Shoot the pig.
- Bleed it.
- Scorch and scrape.
- Scorch it again.
- Wash and scrape.
- Split and gut.
- Split the spine.
- Hang to cool.

Tools and Supplies

- .22 rifle and several long rifle shells. (We prefer hollow points.)
- Two sharp hunting knives, one 5- or 6-inch blade, one 3- or 4-inch blade.
- Old garden hoe.
- Two long flexible dull kitchen knives or patent scrapers.
- Several terry cloth or burlap rags.
- Five or 6 gallons of very hot water.
- Propane "Tiger Torch."
- Meat saw or eight-point crosscut saw.
- Three yards baling twine.

Details

1. Shoot the pig. The place to hit is just above a line drawn

Fig. 3-11. After being cleaned, the carcass is cut into two halves and left to cool.

between the tops of the eyes. Shoot from pig level, i.e., parallel to the spine, not at an angle downward. If you hit the pig square with the .22, it will drop like a stone. Some people like to put a rope on the pig before they shoot so that if they miss it won't run off. We prefer to have the pig in a small pen and shoot from one of the lower fence rails.

2. Bleed the pig. When the pig drops, it will usually lie still for 10 or 15 seconds. Jump over the fence and cut its throat, during these moments. It's hard to describe the spot to cut, but since there are no important cuts of meat in the area it doesn't hurt to overdo it. Stick the knife in just behind the corner of the jaw, then sweep it back and forth on the inside. Cut the esophagus and lots of other things until there's a big gush of blood, then get out of the way fast. As the pig dies, it usually thrashes about violently and can kick hard enough to do serious damage to anything in the way.

3. Scorch and scrape. Turn the torch on the pig just long enough to raise the skin in big blisters. Have your partner use the garden hoe to scrape away these blisters as they form. Do the whole pig in this manner. Don't use a good hoe. The torch will take the temper out of it. A dull edge is best in that it won't cut the skin.

4. Scorch again. This time toast the whole pig the color of very dark toast. Pay particular attention to the hooves. Burn them until they blister, then twist them off.

5. Wash and scrape. Soak the rags in the hot water and lay them on the skin for a few seconds. Throw the rags back in the water and start scraping the spot you have wet. The long flexible knives should peel the char away leaving the skin underneath creamy white. If the char doesn't come off easily, use more hot water. After you've done the first half of the pig, roll it onto some planks to keep it clean. We don't use the head and feet for anything, so we don't worry about getting them too clean. If you're going to make head cheese or pickled pigs' feet, you'll have to clean them well. We find that a rag does a better job than a knife on the hard-to-get-at places around the hooves; just wrap the rag around the foot and twist it back and forth.

6. Split and gut. Once the pig is clean roll it on its back and place a two-by-four along each side to keep it there. If you've left the feet and head dirty, cut them off first. There's a good deal of artistry in gutting an animal properly, but that comes with experience. At first you have to hack away as best you can. There isn't really anything you can do wrong, but it's nicer if you don't spill urine and feces all over the meat. Detailed instructions for this operation are

available from many sources (*Farming for Self-Sufficiency* by John and Sally Seymour is easy to follow). Basically you remove the head, split the chest, abdomen, and pelvis, and pull out the innards. Be sure to salvage the liver, heart, kidneys, and leaf lard.

7. Split the spine. Take your meat saw or eight-point crosscut saw in hand and saw the spine in half lengthwise. It's easier than it sounds. If the saw tends to bind, have your partner hold the pig open for you.

8. Hang to cool. Tie a twine to each back foot and hang the sides from a convenient rafter to cool. Use some cool water to wash away the blood stains and be sure the dog can't get at the meat.

By using the propane torch instead of the traditional scalding tub, you not only save yourself a lot of work, but also save several hours of advance planning. It takes a long time to heat 20 or 30 gallons of water, not to mention rigging a hoist.

Chapter 4

Sheep

If your homestead has a pasture, an idle building, and a fence that will keep out dogs, you may want to consider purchasing sheep. They come in many sizes and produce both meat and wool. In Europe, ewes are also prized for their milking ability. It is from goat and ewe milk that Roquefort cheese is made. Any individual can enjoy raising sheep, whether to eat at home, to market, to grow wool, to produce more lambs, and/or to show at fairs.

A lamb commonly begins life with at least one brother or sister. As a ewe has only two teats, it is often more convenient if she has only two lambs. Otherwise, some will have to be raised artificially. (A relatively recent import, the purebred Finnsheep, produces litters of three to six lambs. Our own domestic breeds can produce triplets.)

The newborn lamb has a long tail that should be removed between five and 10 days of age, if he is healthy. The prevents manure from caking between the tail and body. Caked manure is a source of disease, possible death, and will attract flies in warm weather. If ram lambs are castrated, this is commonly done at the same time the tail is removed and they become wethers. After about four or five months of age, testosterone, the male sex hormone, begins to secrete itself into the meat, giving it a rank odor and taste. This is not a factor if rams are slaughtered before this time. Leaving a ram lamb uncastrated will allow him to grow faster than either a wether or a ewe lamb.

By two weeks of age, lambs are often given the opportunity to eat from a creep. This is a dry, well-lighted place in the barn set apart entirely for their own use with grain, forage, and water. Lambs have access to it through a panel with openings wide enough to allow them to enter, while keeping the ewes out.

At about six to eight weeks of age, all four of the lamb's stomachs will be functional. He can consume and utilize enough feed to make it possible to wean him. Lambs weaned before 90 days of age are considered to be "early weaned" and must be carefully attended.

When they weigh 50 to 60 pounds, lambs being raised for meat should be separated from ewe lambs that will be kept for breeding purposes. In this way, replacement ewe lambs are not exposed to as much grain. This cuts down the likelihood of their becoming overweight. Fat impairs breeding ability and may damage the health of the fetus. Separating the lambs also ensures that ewe lambs will not be inadvertently bred by ram lambs.

If grain fed and pushed, lambs should be ready to eat by the time they are four to five months old. Ewe lambs, who tend to lay on more fat than wethers or ram lambs, may be slower. Ewe lambs saved back for breeding will begin to come into heat at five or six months of age, but should not be bred before they are seven or eight months old. This way, they will lamb as yearlings.

While some people let ewe lambs remain open until they are about two years old to allow them maximum growth, a well-fed ewe lamb will not have her growth impaired to any significant extent by lambing at one year. In addition, she is able to pay for that year's feed in lamb production.

When the ram is turned in with the ewes, his brisket is painted with a special paint. He is fitted with a marking harness so that he leaves a mark on the ewe's rump when he breeds her. Then you can figure her due date. Five months later, when the ewe lambs, the cycle begins again.

When purchasing sheep, you should decide whether you want to raise them primarily as a source of meat or wool. This is necessary as sheep are usually divided into two categories—meat or wool producers.

BLACKFACE AND WHITEFACE BREEDS

Traditionally, the blackface breeds or their crosses (Suffolk, Hampshire, and Oxford are good representatives) and a few whiteface breeds (such as the Dorset and Southdown) are known for their meat-producing abilities. These sheep make very efficient use of their feed, gain rapidly, and produce a carcass with little excess fat. They will yield around 6 to 8 pounds of wool per year, possibly containing some black fiber.

Whiteface sheep and their crosses, including Dorset, Rambouillet, and Corriedale, are sometimes called the mother breeds. This is because these ewes will, in general, take more interest in their lambs than a blackface ewe. They also produce more milk. In addition, they are noted for their wool production, which varies from 8 to 15 pounds of black-fiber-free yield each year.

White-faced ewes have the ability to come into heat year-round (polyestrus) which is an advantage not commonly shared by their black-faced cousins who breed and lamb naturally only once a year.

CROSSBREDS

The crossbred, a product of these two color groupings with a speckled face, is the model of moderation. As a rule, crossbred individuals make good mothers. They claim their lambs well and produce plenty of milk. They yield a moderately heavy fleece, with some possibility of black fiber, and produce efficient, fast-growing lambs if bred to the right kind of ram. Such ewes usually come in heat earlier in the year than a black-face ewe, and they may breed year-round.

Crossbred lambs have a genetic advantage known as *hybrid vigor*. This trait is expressed in a general hardiness and will to live. This is not to say that lambs within the black and whiteface breeds are not vigorous; it's just that crossbreds tend to be more so.

EQUIPMENT

As far as basic equipment goes, there is very little investment required, although you can get as fancy as you like. Sheep require a water bucket or tank, a simple feeder for grain and hay, and a container for salt and minerals. Beyond this, it is a good idea to have a syringe and needles (if you know how to use them), a livestock thermometer, iodine for cuts, and handshears. When catching sheep, and especially when penning a ewe and new lambs together, it is good to have a few lambing panels. These are similar to creep panels except that the slats or bars run horizontally the entire length of the panel.

If you decide that you want to get some ewes or lambs, there are several courses open to you. The best way to get acquainted with the sheep raisers in your area is to ask your neighbors, inquire at the feed store, the vets, or, best of all, visit a sheepman. These people will all be more than willing to acquaint you with who raises what in your area.

Sheep will not fit every homestead. If, however, you find that yours can provide for their few requirements, there is a breed that will suit your needs and more than amply reward your interest and labor.

CHOOSING THE RIGHT BREED

If you are planning to get some sheep, a little investigating can make a lot of difference between being a really happy shepherd and one with all kinds of problems. For one thing, there is a great difference between the different breeds. If you are heavy into spinning you might be happiest with Rambouillets, Merinos, or Corriedales. If you want to produce quality market lambs, you could buy the black-faced crosses. Black-faced refers to Suffolk and Hampshire, or you could choose purebred sheep of those breeds. If you have young children or elderly people in the family who will have the sheep as their special project, you might want to choose one of the smaller breeds such as Southdowns, Dorsets, Cheviots, or Shropshires. There are many more breeds, but these will most likely be the easiest to find. Each breed has an association and can give you information on breeders in your area.

Rambouillets

A fine wool breed, the Rambouillet, came into being in France in 1785, when Louis XVI gained permission from the King of Spain to import 318 ewes and 41 rams from the closely guarded Spanish Merino flock. His purpose was to encourage and improve the woolen manufacturing industry. Further improvement was made in the breed in Germany, especially by Baron F. Von Homeyer, who increased the size and ruggedness of these very fine-wooled sheep. They came to the United States in 1893 and became the backbone of the western range flocks. Later a number of new breeds were formed from crosses of Rambouillet and Merino sheep with other breeds. For some time both breeds had many heavy wrinkles and folds in the skin, but modern breeding has minimized this trait.

"Bullets," as they are nicknamed by some breeders who find Rambouillet a bit formal, are large sheep but quite docile (Fig. 4-1). They are very hardy and easy lambers that generally take good care of their lambs. If you have a range of type setup, these might be the breed for you. They work well with a sheep dog and have a

Fig. 4-1. A Rambouillet is a large, docile sheep.

strong flocking instinct. There are both horned and polled Rambouillet rams. The horns are quite large and very handsome. Select individuals with open faces (no wool around the eyes), long dense fleeces, and large, smooth bodies. The wool can be sold to a professional wool buyer, direct to a factory, or to home spinners.

Corriedales

Corriedales are similar to Rambouillet. The breed was developed in New Zealand and Australia and is a combination of the Merino and the Lincoln, a long wool breed. The Lincoln is very large and has a course fleece that often grows to over 12 inches in length in a year's time. Fleeces will weigh 15 or 20 pounds. The Corriedale reflects much of the value of the Lincoln in that it has improved meat characteristics. The fleece is of medium fineness and is long, bright, and dense. Corriedale fleeces sell very well, and the carcasses of the lambs are good. Much of the commercial lamb from New Zealand imported to the markets of the United States are from these ewes.

Hampshires

Hampshire sheep are big boned, muscular sheep with dark

faces and legs, but having some wool covering on both (Fig. 4-2). They are of quiet temperament and are good mothers. Excellent fat lambs for show and commercial market can be produced from the Hampshire. Yearling rams can be sold to commercial range operators for use on the white-faced ewes, such as Rambouillet and Corriedale, to improve the meat conformation of the lambs. The lambs are fast growing and often gain a pound a day when on a fattening ration, although they are very good gainers on just grass and mother's milk. The meat is high quality, which some attribute to inheritance from the Southdown.

The Hampshire was developed in England from bloodlines of the Wiltshire, Coltswold, Berkshire Knot, and the Southdown. Old records show that there were several flocks in this country in 1840, but they disappeared during the Civil War (perhaps to feed hungry soldiers). The sheep reappeared in 1880, and one account tells of a huge ram weighing 200 pounds being imported to a Mr. Metcalf. Today Hampshire rams range from 250 to well more than 300 pounds.

Wool is of medium texture, but the fleeces are not the main consideration in selecting Hampshires as a breed, as they will

Fig. 4-2. A Hampshire is a muscular sheep.

weigh around 8 pounds or so. While the Hamp lambs will eat about 30 percent more feed to reach market quality than the smaller Southdown, the lamb will be at least 30 percent heavier. In selecting stock look for good size and open faces. The more upstanding and higher off the ground types are favored in the show ring at the state and county fairs.

Suffolks

Suffolk sheep are enjoying great popularity at this time. A large, hardy breed with black heads and legs that are clear of any wool, Suffolk sheep are quite aggressive foragers and the lambs are very fast gaining. They do not have the tendency to put on fat that some of the other breeds do. As a result, they can be slaughtered at 120 to 150 pounds, yielding a 60 to 75-pound carcass or more that is quite lean but well-finished. They are in great demand in the show ring for both market and breeding classes and command higher prices than other breeds of sheep in most areas of the country at this time.

Originating in Suffolk County in England, they were formed by crossing a large strain of Southdown with the Norfolk. The Norfolk was a long legged, active sheep with horns and black faces whose fleece was rather light. They were noted for being highly prolific animals and through selection the horns gradually disappeared, although today some lambs will have small scurs that are easily removable. Every so often you will see a ram that is a throwback with short horns. The long narrow heads and the shape of the shoulders are an aid to easy lambing, and the lambs are generally hardy and active. The fleeces are of medium fineness. Black fibers in the wool are to be avoided. You can sell rams to range sheepman, lambs for market, and show lambs and breeding stock for show or for others' flock needs.

Dorsets

The Dorset used to be one of the smaller breeds, but the association is very progressive and has been promoting longer legged, larger bodied sheep for some time now. Dorsets are all white with a wool texture that many spinners like to work with (Fig. 4-3). There are both horned and polled Dorsets, and they are quite docile sheep with what might be described as a "Mona Lisa smile." One of their most outstanding points is that they will lamb at almost any time of the year, which would enable you to sell lambs when the price would be highest in your area. They have useful meat charac-

Fig. 4-3. A Dorset sheep is totally white.

teristics and are heavy milking ewes that are good mothers. It is interesting that in some European countries the Dorset has been used as a dairy animal, for both milk and cheesemaking. They are quite popular in the show ring.

Cheviots

On the green rolling hills of old English estates some 200 years ago, one could see, while riding by in one's carriage, flocks of small white puffs of wool with clean white heads and legs and alert pointed ears and prominent dark eyes (Fig. 4-4). These stylish sheep were called Cheviots and are still popular today in the United States. They are very active sheep and withstand cold weather better than some other breeds. Cheviot sheep seem to do especially well on hilly areas. They do not have the strong flocking instinct of the fine wool breeds but these hardy little sheep will return to the barn from grazing in the woods so they can be protected from the bears and bobcats at night.

Shropshires

Another breed that owes its origin in part to the Southdown is the Shropshire. A little smaller than the Hampshire, it has wool on the head and legs, but selection is for open faced, longer legged sheep (Fig. 4-5). While not quite as common as they once were, The Shrops are useful sheep with good carcasses and a medium wool. They are seen both as farm flocks and in the show ring.

Southdowns

One of the smallest breeds today, but the breed that has been used extensively to form other breeds because of the outstanding meat qualities, is the Southdown. For many years the champion carcass lambs at the livestock shows were almost always Southdowns. Then the larger breeds began to gain in favor in an effort to produce a carcass with a larger loin eye and heavier weight with less fat covering. They are slower growing than some other breeds, but they also eat less. Once they have reached maturity, they maintain good condition on little feed. The fleece is very light in weight and does grow exceptionally long. The wool extends to the knees and hocks and over the head, and some wool cover is on the face. Legs, face, and ears are light brown or gray. This breed is nice for small children to work with as they are very docile and easily managed. The lambs are especially appealing.

There are around 25 breeds raised in the United States. Each has its own characteristics and strong points and weaknesses that

Fig. 4-4. A Cheviot sheep has pointed ears and dark eyes.

Fig. 4-5. A Shropshire sheep is a little smaller than the Hampshire.

the breed associations are attempting to improve on.

Sheep are especially adapted to homesteading as they give so much in return for the little that they require. They need forage, clean water, mineralized salt, a protected place to go in very stormy or hot weather, and they need to be protected against dogs or wild predators. They can give you meat, wool for clothing, rugs, pillows and comforters, candles, milk, cheese, soap, lanolin, rennet, paper, violin strings, drum heads, and saddle pads. Many other valuable items to make life better for man are manufactured from sheep products.

If you are planning to go down the road and select some sheep from a nearby farm, it might be wise to take along someone who has some knowledge of them. If you would like to raise sheep that could be exhibited at the county and state fairs and be sold to other sheep enthusiasts for a higher income on your investment, you might like to contact some of the breed associations for further information and a list of breeders near you.

American Cheviot Sheep Society, Rt. 2, Lebannon, VA 24266.
American Corriedale Assn., Seneca, IL 61360.
American Hampshire Society, Rt. 10, Box 199, Columbia, MO 65201.
American Rambouillet Sheep Breeders Assn., 2709 Sherwood

Way, San Angelo, TX 76901.

American Shropshire Registry, P.O. Box 1970. Montecello, IL 61856.

American Southdown Breeders Assn., Rt. 4, Box 14 B, Bellefonte, PA 16823.

American Suffolk Sheep Society, 55 E. 100 N., Logan, UT 84321.

Continental Dorset Club Box 577, Hudson, IA 50643.

National Suffolk Sheep Assn., Box 324 S, Columbia, MO 65201.

Barbados Blackbelly

The Barbados Blackbelly, a tropical hair sheep, is originally from the Island of Barbados in the West Indies. Though introduced into the United States only in 1904, it now numbers well more than 500,000 in the United States. The breed's unique characteristics account for much of its popularity and make it ideally suited for the smaller farmer or homesteader.

Among these special qualities are the ability to lamb throughout the year, with a lambing interval of approximately seven months, and a high percent of multiple births, averaging 1.78 lambs/lambing. Thus the Barbados can be raised along or crossed on domestic breeds to increase productivity.

The Barbados Blackbelly also comes in a variety of unusual colors, ranging from the base black and tan color (much like the color of a bay horse) through yellow, tan, black, and pinto colors. These combinations make for a colorful flock. Because they are a hair sheep, when wool is present in the hair coat, it tends to be more

Fig. 4-6. A yearling Barbados ram with horns and cape typical of the breed (photo by Paddy O'Reilly).

Fig. 4-7. A Barbados ewe and two lambs (photo by Paddy O'Reilly).

visible in the fall and winter and then shed in the spring.

The mature Barbados ram is a particularly impressive-looking creature. Weighing 125-135 pounds, the rams have a piece of long hair, called a cape, on the underside of the neck that extends down into the brisket and reaches full development in the fall (Fig. 4-6). In some rams this cape extends over the sides of the neck and shoulders as a blanket. Horn development in the Barbados ram in this country is variable, but increasingly they are being selected for a horn that is similar to that of the Bighorn sheep or the Texas Longhorn. The ewes occasionally have very small horns, but this is unusual.

Additional attractive features of the breed are their resistance to internal parasites and their ability to withstand cold or heat well. They are particular alert and intelligent as compared with other sheep and make excellent pets. Ewes with very young lambs are highly protective of them, and some rams will charge dogs and other potential predators, which can help reduce losses from dogs or coyotes (Fig. 4-7).

This remarkable breed has many different potential uses. It can be raised as a meat breed; the Barbados has considerably less body fat than other comparable sheep, and the meat tastes more like veal than mutton. It can be crossed with domestic breeds of sheep, such as the Dorset or Suffolk, to increase producitivity. The limited estrus cycle of the Suffolk breed is very dominant, so they don't tend to cross as well with the Barbados when selecting for year-round lambing ability.

Rams of some of these flocks, ferally reared, can be used for hunting purposes, which is a potential market many sheep breeders may not have considered. Because of their rare colors, horn development, and antelopelike running and leaping ability, many breeding flocks are being established by game management officials. In California a few small breeding flocks have already been established, primarily for hunting. Particularly outstanding Barb rams are bringing $200-$300 for stud purposes in these game ranch flocks.

All in all, the Barbados is a unique breed, with a great deal of, as yet, unexplored potential for the small farmer or homesteader. Whether you raise them for pets, to provide yourself with a good supply of lean meat, or for market purposes, the Barbados is sure to make your farm a more interesting place.

FINNSHEEP

The Finnsheep Breeders Association titles Finns "the high fertility breed" and with good reason. Unlike our conventional domestic breeds, the Finnsheep is celebrated for her inherited ability to conceive and give birth to litters of up to six or seven lambs.

American sheep producers, recognizing the potential for increased profits, enthusiastically bought up breeding stock after it became available in 1966. It soon became apparent, though, that the litter-lamb concept was not the answer to all profit problems.

Now that Finnsheep have been in this country long enough to provide some practical on-the-farm experience, the place of Finnsheep in both commercial and purebred operations can be evaluated with some degree of reliability. Whether in a crossbred or purebred operation, the major factors to evaluate when dealing with Finnsheep include prolificacy, cost of raising to market weight, and carcass evaluation.

Some wild claims and resulting high expectations have been generated by secondhand reports of producers achieving lambing percentages of 400 to 600 with crossbred Finns. In reality, the Finn cross ewe can be expected to produce a "one percent increase in percent lamb crop weaned for each percent. Finnsheep breeding in the ewe", according to the Finnsheep Breeder Association, Inc. of Minnesota. Therefore, a flock of ewes that are 50 percent Finn can be expected to wean 50 percent more lambs than during the previous lambing. If the ewes had a 100 percent lamb crop weaned, but now uniformly contain 50 percent Finn blood, they can be expected to wean 150 lambs per 100 ewes.

In University of Minnesota trials, half-Finn yearling ewes had a lamb crop of 138 percent and a 208 percent lamb crop born at three years of age. It is clear then, that percent lamb crop born and weaned depends not only on breed, but on age of ewes, culling intensity, nutrition, and management.

An important factor in the Finn's prolificacy totals for her lifetime is the ability to consistently conceive at an earlier age than most of our domestic breeds. In any sheep operation, little income can be generated from a ewe that fails to conceive and lamb. In most cases today, breeding ewes to lamb at a year of age is certainly economically advisable, and the chances of her setting are higher with the introduction of Finn blood.

Because a Finn is a whiteface breed, it stands to reason that crossing her to another whiteface breed should do much to insure a polyestrus offspring. When crossed a blackface breed, results can be expected to parallel those of any black-whiteface cross, i.e., some increase in length of breeding season, but no definite tendency to breed out of season. With Finnsheep the critical factors involved are cost of artificial rearing of extra lambs or bonus babies and rate of gain.

By experimenting with different degrees of Finn blood, it is eventually possible to achieve an almost perfect balance of 200 percent, 250 percent, or whatever number of lambs the ewes can successfully raise by themselves. This can be accomplished in a relatively short length of time with Finn crosses. As with our domestic breeds, there will be incidences of orphaned lambs. Where Finns are concerned, there will be an increased number of lambs that, for one or another reason, cannot be mothered naturally and must be raised artificially.

Methods of raising these surplus lambs are as varied as the situations from which they arise, and it is clear that no one method will best fit every situation. Many producers have found it not only advisable, but necessary, to set up a functioning lamb nursery that will allow extra lambs to be fed and managed with minimal labor. Lamb nurseries range in variety from setting aside a light, draft-free area of the barn as a pen in which to bottle fed lambs, to the construction of a building specifically designed as a lamb nursery unit complete with automatic milk mixing, feeding, and cleanup devices.

In general, cost factors in raising extra lambs include milk replacer cost, equipment, time available, labor involved, and skill on the part of the manager in the area of orphan lamb raising. Finn lambs tend to be very vigorous and hardy at birth. This factor makes it easier for the extra lambs to survive the first critical 24 hours of adjustment to artificial rearing.

Table 4-1. Comparison of Weights of Various Breeds.

Breed of Sire	No. of Lambs†	Birth	Age in Weeks				
			10	14	18	22	26
Rambouillet	77(36)	11.0 lbs.	49.9 lbs.	61.1 lbs.	90.1 lbs.	112.2 lbs.	133.4 lbs.
Dorset	75(27)	11.2	50.6	66.9	93.8	115.8	137.2
Coarse wool	74(36)	11.0	51.4	64.8	90.4	111.8	133.1
Finnsheep	143(59)	10.0	49.7	65.4	88.1	105.8	122.6

†The lambs were slaughtered at either 22 or 26 weeks of age. The number slaughtered at 26 weeks is shown in parenthesis.

Once the lambs reach a point where they are weaned from the ewe and/or milk replacer, the main concern is to get them grown to market or breeding weight as cheaply and quickly as possible. Whether this involves pasture, grain feeding, or both, it is clear that Finnsheep grow more slowly than our conventional meat breeds, whether as a purebred or a crossbred. For this reason, many who raise Finnsheep primarily for slaughter find it best to cross Finns with a meat breed that is well proven in rate of gain and feed efficiency.

As is apparent by the following study done in 1970, by the Clay Center, Nebraska, U.S. Meat Animal Research Center (Table 4-1), even crossbreeding Finns will not necessarily result in individuals who are leaders in rate of gain.

Unfortunately, no representative of leaders in meat production, such as the Suffolk, was tested. The Dorset can be considered a representative of the meat group, although it is somewhat more of a multipurpose breed.

In spite of the fact that Finnsheep and their crosses are not leaders in rate of gain and feed efficiency. Finn crosses stand up well in carcass evaluation, depending on the individual. As a crossbred containing 25 percent and even 50 percent Finn blood at times, it is not possible to distinguish the Finn in appearance from any common black/whiteface cross.

Table 4-2. Carcass Evaluation of Various Breeds.

Breed of Sire	Carcass Weight	Carcass Grade†	Rib Eye Area	Fat Thickness	Kidney Fat	Est. Yield Trimmed Retail Cuts
	lbs.		inches	inches	lbs.	lbs.
Rambouillet	65.2	5.4.	2.26	0.19	2.7	47.3
Dorset	69.0	6.5	2.41	0.18	2.9	49.5
Coarse wool	67.0	5.3	2.28	0.20	3.0	48.1
Finnsheep	63.2	5.0	2.13	0.20	3.1	45.4

†Carcass Grade: 3=high good; 4= low choice; 5= average choice; 6= high choice; etc.

Again, results from Clay Center tell the story (Table 4-2). Although the Finn does not outshine the others in this comparison, the figures they present make the early titling of "meatless wonder" by some sheep producers unfounded.

From the above consideration of the Finn's prolificacy, rate of gain, and carcass evaluation, it is clear that this breed has a definite place within today's sheep industry where an increased number of lambs from a minimal number of ewes is desired. To achieve this goal, her greatest potential would seem to lie in using the breed as a crossbred mother. In this capacity, the Finn would be passing on her best but tempered qualities, such as early maturity, prolificacy, and mothering ability, while giving the other half of the cross a chance to increase the weak rate of gain and efficiency factors yet still improving overall carcass quality.

FENCING ALTERNATIVES

Fencing nowadays is an expensive proposition. Because it is a major investment, consider all alternatives and make your plans carefully.

Ordinary barbed wire fences will not hold sheep. The spaces between the strands are too large, and the wire stretches and sags easily. It is possible, if you have an existing barbed wire fence in excellent condition, to add more strands and to make a point of keeping it in optimum repair. It is not a happy solution.

The usual electric fence will contain only trained sheep in a contented state. They must be trained by taking them to the wire without letting them stampede through it, tempting them to reach for grain or other treat, thereby touching their uninsulated noses and ears to the current. They can be trained fairly easily, in small groups, and will respect the fence under most conditions. Excited sheep will demolish the setup, whether they're trained or not.

The most inexpensive woven wire fence, 26 inches high with 12-inch stays, will contain most sheep. Unless you have some "juniors" or a nervous breed, they will not try to go over it, even though they are all capable of doing so. The large squares prevent the sheep from getting their heads caught, but their curiosity and appetite for greener grass will put a lot of stress on the fastenings. Lambs can easily get through the lower openings. Further, dogs and coyotes will have no trouble getting through or over such a fence. Therefore it's more suitable for adult sheep, and for subdividing pastures, rather than bounding them.

The ideal fence is made of woven wire at least 36 inches high, with 6-inch stays. These should have an additional barbed wire strand below it, and two or more above it. Wooden posts for this type of fence need to be at least 5 inches in diameter, of suitable wood, preferably treated, and sunk 2-3 feet deep. Steel posts can also be used. The fence must be strung tight, and the end posts reinforced. There is no point in spending great amounts of time and money on a fence if it is going to sag and be useless in a few seasons.

Another very satisfactory fence is the New Zealand type of electric fence. This is a high, multistrand, arrangement, with a special high-powered charger, which will not short-circuit even when overgrown by weeds or wet brush. Posts need to be driven only at very long intervals (up to one-half mile), with light fiberglass posts and line spacers supporting the wires in between. Tension is maintained with springs at the heavy-duty strainer posts. Materials and labor for erecting this fence are far less expensive than for the traditional page wire fence. It has proven predator-proof as well.

Wooden fences are perfectly satisfactory if the horizontals are close enough together and if they are firmly joined. They would be too expensive for most purposes, but make a fine attractive paddock.

For rotating pasture, the new electric net fencing should be considered. It is not cheap, costing twice as much as page wire fence, but it includes its own posts and is far more convenient to erect than any other type. Several hundred yards can be carried easily by hand, and the light posts just stick into the ground.

So-called "snow fence," wooden roll fencing, is also expensive but convenient to erect for temporary purposes. It will follow contours of the land, will effectively discourage dogs, and will prevent sheep from getting interested in the greener grass since there is no opportunity for them to stick their heads through. Fiberglass, steel, or thin wooden posts can be erected at long intervals and the fence wired or tied to them.

The cheapest barrier is simply woodland, for sheep will not run off into the woods as cows will and don't require any fence to keep them out of the neighbor's woodlot. Usually the main purpose of sheep fencing is to keep predators out.

FEEDING RACK

A feeding rack for both hay and grains allows the consistent feeding necessary to produce a wool clip of good weight and quality. This hopper design has two advantages: keeping the sheep's heads out of the hay and preventing chaff and seeds from getting into the

Table 4-3. Materials List for the Sheep Feeding Rack.

No. of Pieces	Length	Dimensions	Use	Hardware
6	3'7"	1¾" x 1¾"	posts	1 pound 6d nails
3	3'6"	⅞" x 6"	bases	2 pounds 8d nails
2	3'6"	⅞" x 10"	ends	
2	3'	⅞" x 3"	ends	
2	2'10"	⅞" x 3"	ends	
2	2'8"	⅞" x 3"	ends	
2	2'6"	⅞" x 6"	ends	
3	2'3"	⅞" x 2"	ends of trough	
2	11'10¼"	⅞" x 7"	sides of trough	
1	11'10¼"	⅞" x 6"	trough partition	
2	12'	⅞" x 10"	trough bottom	
2	11'10¼"	⅞" x 10"	top boards	
46	2'2½"	⅞" x 3"	slats	
6	10"	1¾" x 3½"	furring to nail top boards	

wool. The trough catches the hay leaves so they may be consumed instead of falling into the bedding.

Make the rack longer or shorter to suit your flock's needs. If you make it less than 8 feet, omit the two middle posts and their cross ties. See Table 4-3.

Use the 8d nails at places where nails are driven into the 1¾-inch pieces. The 6d nails are used at all other points. The lower ends of the slats are beveled and nailed to the bottom edge of the trough partition (Figs. 4-8 and 4-9).

EATING HABITS

Anyone who has kept a flock of sheep will agree that they are creatures of habit. Sheep, like other animals, develop interesting personality habits if they are around humans consistently.

Many a study has been made on ovine behavior over the years, and one thing they agree on is that sheep behavior varies depending on the geographical location, the degree of nutrition, the breed of sheep, the climate, and the season. For instance, the Cheviot, Columbia, Suffolk, and Rambouillets are strong foragers and will cover a lot more distance on a large pasture than the Hampshire, Southdown, or Romney Marsh. This is especially true in steep hilly country. While some seem to be content to stay in one small area when they find the grasses especially tasty, others seem to have a strong urge to keep on the move, always looking for a greener pasture. Doesn't this remind you of the ewe that always has her head caught in the fence because she has spied one irresistable

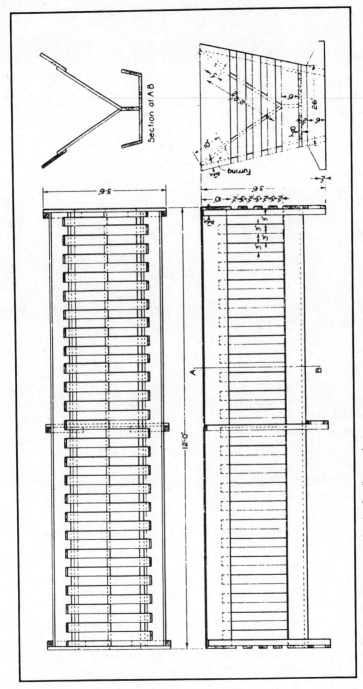

Fig. 4-8. Construction details for the feeding rack.

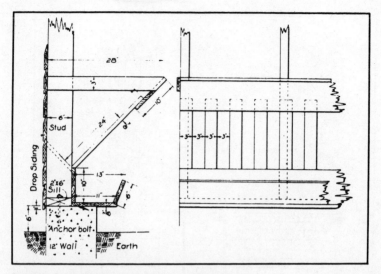

Fig. 4-9. Construction details for a wall feeding rack.

mouthful in the next pasture even though the same plants are growing in her own area?

Grazing time is from daybreak through the early morning, then it is time to "shape up" and chew the cud until supper time—late afternoon until dusk. The setting sun signals them to seek a safe place, often along the fence nearest the barn, or in the barn or barnyard. On the range they will seek out a high place. The grazing periods last about 10 hours out of every 24 and are broken up into from four to as many as 15 different occasions. Ruminating may go on as long as 10 total hours in 24, and each cud may be chewed from anywhere from a minute to an hour or so.

Studies of ruminants with fistulas in their sides (a permanent opening through which the action and contents of the stomach compartments can be examined) indicate that the rumen and recticulum are contracting about once every minute whether the animal is eating, ruminating, or resting. This action only stops after a three or four day fast.

Much has been said about the sheep's habit of eating plants down close to the ground. This has been found to exist when the eating area is restricted and they are not permitted to move on. They are fastidious diners and avoid eating grasses that have been soiled by feces and any hairy or greasy plants. Several studies cited the evidence that sheep are health food conscious. It was found that they will often choose the foods that will give them a balanced ration, or correct a particular deficiency that they may be suffering from, and shun junk foods. They will also have favorite recipes and consistently choose a certain preparation when offered that same food in several different forms.

Lambs' eating habits are rapidly developed from a few moments after birth. Nearly all lambs will have found the teats and mastered the pushing up on the udder and sucking the teat within 10 minutes to one hour after birth, unless they are very weak. It is important that the newborn lamb get some warm milk into its stomach within an hour or so after birth, or it gradually loses the sucking instinct and must be helped by humans if it is to survive.

A group of lambs may nurse as many as 65 times in a given 24-hour period in the first week of birth. Later the mother cuts down the frequency considerably and will only stand for a period of 10 to 20 seconds at a time for nursing. If she has twins, after identifying one baby by a sniff of the tail (which, by the way, is in ecstatic motion), she will often call the other lamb to her side for a snack.

The lamb will be eating some solid food by two weeks of age. If the lamb is offered solid food in a creep, the rumen will develop and come into use shortly thereafter.

WORMING

Internal parasites can be a serious problem in sheep, especially in lambs and older individuals. Therefore, it is necessary to establish a routine worming schedule.

An infected sheep often appears thin and produces an inferior fleece. In some cases, diarrhea and extreme thirst are symptoms of internal parasites. Death results from severe infestations.

To prevent a buildup of internal parasites, healthy sheep are generally wormed three or four times a year before being put on pasture, when the lambs are weaned and before being confined for the winter. It is best to worm a few days before a move to new quarters, as worms are then left behind. Every producer must establish a schedule that will best fit his sheep's needs.

Anthelmintics can be administered in several ways—by drenching, through feed additives, by injection (in some cases), or in pill form. In the majority of cases, drenching is the most reliable, effective, and least expensive method. Unless it is done properly, drenching also offers the greatest opportunity for injury to the sheep through introduction of wormer into the lungs.

The epiglottis provides a "gate" in the sheep's throat, which closes the entrance to the windpipe while the sheep is swallowing,

Fig. 4-10. Place the sheep's head in the correct position for drenching. This will help avoid accidents.

or the liquid is forced down the throat too fast, the gate to the windpipe doesn't close, and the liquid enters the lungs. This can cause at the least choking and, very possibly, pneumonia and/or drowning.

For this reason, it is important that a few simple steps be followed when drenching. Pen the sheep either in a chute or a corner of the barn. There need only be room to allow the operator into the pen and to prevent the sheep from overcrowding and injuring one another. It is always a good idea when handling livestock to drench during the cool hours in order to avoid unnecessary stress from overheating.

If drenching is done with sheep in a chute, position yourself just behind the sheep's shoulder. If they are in a pen, straddle the sheep just behind the shoulders, where you can control their movements by squeezing with your legs. If the sheep is large and you are small, it may be necessary to back it into a wall or panel, as sheep invariably back up when the drench gun is inserted.

Fig. 4-11. Insert the drenching gun about this far.

Fig. 4-12. The gun is fully inserted, and the plunger is depressed.

Before you begin, be sure the sheep is standing and her head and neck aren't twisted. Then place one hand under her jaw to keep the head in a natural position for swallowing (Fig. 4-10). Gently insert the drench gun along the side of the mouth over the teeth and tongue (Fig. 4-11). The nose of the gun need not go in the entire length, depending on its size, but should stop at the bulge of the tongue at the rear of the mouth. Point the barrel toward the side of the mouth rather than straight down the gullet. Take care not to elevate the sheep's head unnaturally high. The nose must never be higher than the eyes, as this impedes swallowing and proper closure of the windpipe.

Depress the plunger slowly, so the sheep can swallow with no difficulty (Fig. 4-12). If she begins to choke, immediately remove the gun and let her head drop so she can clear her lungs of any small amount of fluid. Never pull the tongue out of the mouth or impede it in any way, as it is needed for swallowing.

In late pregnancy, the stress of drenching may cause abortions in some cases and lambing paralysis in thin ewes. Establish a routine that avoids this.

Loxon, Thiabendazole, Tramisol, and Phenothiazine are common wormers. Most producers find that rotating brands results in the widest spectrum worm control.

Phenothiazine will lead arsenate should never be used as a drench for pregnant ewes because it causes abortion. Also, if Phenothiazine drips on the wool, it results in a stain that cannot be scoured out. In addition, some sheep have adverse reactions to a combination of Phenothiazine and sunlight (at times fatal). Shade must be provided for several days after its use.

It is a good practice to use a liquid wormer as soon as possible after it is mixed. Some wormers will lose some of their effectiveness within a couple of weeks. Keep the drench gun clean. Check it frequently to be sure no sharp edges develop. As each sheep is wormed, mark her with chalk to prevent accidental overdose.

HOOF TRIMMING

Foot trimming is even easier than drenching and can be done at the same time. It is a good practice to routinely check feet three or four times a year. This is especially true with white-faced sheep, whose feet tend to grow very fast, and sheep living in wet conditions, an environment which favors development of such problems as foot rot.

The only equipment needed is foot trimming shears, a jackknife, or a hoof knife as is used for horses. Whatever you choose, be sure it is sharp.

For this operation, it is necessary to put the sheep in a sitting position, leaning her weight back onto your legs as you bend over her to trim her feet. To get into this position, there is a method that can be used effectively by almost any size shepherd. Place your left hand under the sheep's jaw, turn her head to the right. At the same time, place your right hand under her right rear flank and lift. As you continue to quickly bring her head toward her flank, pivot her rear toward the ground. She will fall onto her rump.

Inspect each of the feet in turn. Trim or cut away any overgrowth of the horn or softer sole until the hoofs are level (Fig. 4-13). Until you learn how deeply to cut, trim carefully. If much bleeding occurs, apply a pressure bandage until it stops and treat with a wound disinfectant. Proper inspection, sanitation, and trimming will prevent most foot problems in healthy sheep.

Fig. 4-13. This sheep looks pretty comfortable as she gets her hooves trimmed.

STOPPING FOOT ROT

Foot rot can be an annoying disease problem in a sheep flock. It makes the sheep lame and interferes with their ability to breed. Ewes produce less milk for the lambs, and the lambs don't grow well. It takes time and labor to treat infected sheep.

Foot rot can be eliminated from the flock with work. The responsible bacteria can not survive long in the environment; it lives only on the hoof wall. Your farm can't be contaminated with the problem, only the sheep are contaminated. If you treat every case properly and then cull out those sheep that have chronic problems, you can get rid of foot rot.

The infection typically starts out with flaky, discolored skin between the toes. This eventually spreads to the heel, undermines the sole, and finally affects the hoof wall itself. The hoof becomes enlarged, and the toes grow unusually long. The foot frequently feels hot where the skin meets the hoof.

Proper treatment starts with a good trimming. Infected hooves should be trimmed thoroughly and all the dead excess hoof tissue removed. It is helpful to get air to the infected parts.

Then the hoof should be submersed in one of various drug or chemical treatments. A 10 percent formalin solution is quite effective. Some sheepmen use a solution of tetracycline or chloramphenical in alcohol. This treatment should be repeated a week to 10 days later.

Injectable antibiotics, such as penicillin-streptomycin, used for several days when the hooves are treated can help assure successful cures.

When they are treated, the sheep should be kept in a dry place for at least 24 hours. If they are turned onto a wet pasture or muddy barnlot, the treatment is less likely to be effective.

Foot rot is introduced into a herd and maintained and spread by carrier sheep. The disease can be eliminated by eliminating the carrier sheep. The flock can be kept clean by making sure you don't bring new carrier sheep into the flock.

Goats and deer can develop foot rot of the type that affects sheep, so they might transmit the disease to sheep. Foot rot in cattle is a different disease and cattle are not usually considered a source of infection for sheep. Still, cattle infected with foot rot should not be allowed to mingle with a clean flock of sheep.

Some sheep are chronic carriers of foot rot. Either they will not respond to treatment or else they will have recurring cases of the disease. These sheep will be a source of reinfection for other sheep in the flock, and they should be culled from the flock.

FINDING THE RIGHT RAM

Select your ram. Don't buy him because he's cheap or convenient. Half the qualities of your lambs are going to be his. This one animal will affect your flock more than 10 or 20 ewes will, so he's worth your time and money (Fig. 4-14).

Make sure he breeds. If possible, buy a proven sire and see some of his progeny. Otherwise, get a guarantee that he will breed that season. Use a marking harness and change the crayon every 16 days. If the ewes are marked a second and third time, get a new stud. You lose a lot of effort and money on a bunch of unbred ewes.

Rams on the show circuit are usually overfed and overfat and often temporarily useless. High heat and humidity can make any ram temporarily sterile. Give rams plenty of shade, exercise, salt, and water.

If you have a problem with wool blindness, get an open-faced buck who breeds true. In five months, you will have a group of open-faced ewe lamb replacements. If your fleeces are light, choose a ram to improve your clip. If your lambs take too long to get to market weight, make a point of finding a sire with good growth rate.

Fig. 4-14. The ram is half the flock; select him with care (photo by Jean Martin).

Keep any new animal separate for a few weeks, trim and examine hooves, worm, and treat for ticks and lice if necessary. Just as the ram can single-handedly bring size and vigor to your flock, he can also introduce foot rot and internal and external parasites.

Don't buy a ram at a public livestock auction if you don't enjoy gambling. Purebred sheep sales and auctions, where the seller is identified and provides guaranteed stock, can be a reliable source. It's best to find a reputable sheep raiser whose operation you can observe. Pay his price: it will be reasonable.

No single ewe should be worth as much to you as your ram. He will cost more, take more care in selection, and be harder to replace because he is, after all, half your flock.

LENGTH OF DAY AFFECTS BREEDING

If you talk to sheep raisers about what time of year to breed your ewes, they'll all tell you that the usual breeding season is in the fall (at least in the Northern Hemisphere). Almost all of them will also tell you that the rams and ewes are made ready for breeding by cool weather.

The fact is that temperature has nothing at all to do with it. It has no effect on bringing on the start of estrus or heat in ewes, nor does it have any influence on the ram's physiological ability to impregnate the ewes. The only temperature effect is that the ram may tire more easily in hot weather than in the coolness of autumn, and that the number of live sperm may be reduced by extreme heat in some situations.

The question was extensively studied by a researcher named Yeats who, 30 years ago, presented his work. It showed that it was the length of the daylight period that influenced the onset of estrus in ewes and sexual activity in rams.

Some common sense things tell us that it is length of day (photoperiod) that is the factor, not temperature. A couple of examples should suffice.

In the mountainous West of the United States, and elsewhere in the world, sheep are commonly trailed or hauled by truck to the high mountains in summertime to take advantage of the fresh, green pastures when the lowlands dry up. We've seen many flocks of sheep in the mountains of Wyoming and Colorado in summer when nighttime temperatures are below freezing nearly every night. The ewes don't come into heat.

In heated buildings or in the subtropics ewes come into heat just as if they were in a more temperature region, even though temperatures don't drop much even at night. In equatorial regions ewes will come into heat almost any time of year. Why? The length of daylight is the same year around near the equator, and ewes seem to be able to cycle any time under such conditions. It is in temperature regions where there is a clearly defined breeding and nonbreeding season.

Remember that the closer one is to the pole (either North or South), the greater is the seasonal difference in day length. Within the borders of the United States alone we can discern differences in latitude and effects.

Two groups of ewes, one in Idaho and another in Texas, were compared each month of the year to check how many were in heat. The ewes in Idaho showed a marked seasonalism with only 2 percent in estrus in May and 100 percent in November through February. The ewes in Texas had 31 percent in estrus in May and 10 percent in November through February. The Idaho area, near Dubois, has cool nights and even frosts in May, when estrus was at a minimum.

These photoperiodic factors also influence sexual activity in rams. Researchers have found that sperm production is at a maximum after summer, increasing with decreasing day length. Also, sperm abnormalities decrease with decreasing photoperiod, so the total quality of the sperm is high in fall.

Studies of testicle size (a measure of sperm and secretion production) have shown that size is small from January to May, increases rapidly in June and July, and reaches a maximum in early August. Regression in size occurs most rapidly during the short days of late November and December. An observant shepherd will have noticed this seasonal size change without need for detailed measurements.

Animal physiologists have also designed experiments to test the photoperiod theory. For his doctoral thesis at the University of Paris, one graduate student confined a group of rams indoors and, for two years, completely reversed the lengths of light and dark periods compared to another group of rams kept outdoors. He found that the annual change in testicle size was also totally reversed. Another researcher confined a group of rams and arranged the lighting to stimulate the whole year's change in only six months. The testicles changed in size in the usual way, but twice as fast.

With ewes, sexual activity is similarly controlled. A scientist named Thwaite did confinement experiments that reversed the usual breeding and nonbreeding seasons by light control. The studies of ewes have been more recent because there is no way as simple as measuring testicle size to judge sexual activity quantitatively. Modern studies with ewes have the advantage of accurate, rapid analytical techniques for measuring hormone levels and the like.

Ewes can be brought into heat by hormone injections or by the presence of a ram. Either stimulus works better in the fall and less well in the spring because ewes are preconditioned by changes in photoperiod. Incidentally, ewes will come into heat about 15 to 20 days after exposure to a ram, and a vasectomized teaser ram can be used to synchronize heat cycles in that way.

A few sheep breeders never did fall for the temperature thing. They were convinced that it is the short days of fall that trigger the onset of ovulation. Alas, even these savants are apparently wrong. Artificial lighting studies have shown that it is the long days of summer that predispose the ewe to come into heat, with a lag time of about 10 to 25 weeks (depending on breed) before estrus begins, coinciding with the normal breeding season.

Some of you are probably already thinking of tinkering with your barn or shed lighting to change your flock's breeding season. It's been tried and does work.

The trouble is that one needs a lightproof building to change seasons radically, and that means enormous increases in expense because of the costs of lighting and ventilation among other things. Happily, the fact that long days, not short ones, are the trigger offers a solution to the problem. Ortavant, a French physiologist, experimented with light cycle cues to sheep and found that a long day could be simulated by giving an extra hour of light during the seventeenth hour after sunrise.

This scheme has the decided advantage of not requiring a light-tight building and consuming little electricity for lighting. An outdoor area could presumably be used provided the lights were bright enough. If you're interested in breeding your ewes out of the usual season, get yourself some lights and go to it. Remember that you can't use a clock timer because it is the period 16 hours to 17 hours after sunrise that counts. Sunrise time changes from day to day.

If you're going to try this, you should also be advised that studies have shown that ewes gaining weight in late winter and spring will come into spring heat more readily. Don't expect winter-starved sheep to respond very well—spring flush them as it were. A ewe will breed more readily if she is in good health and on a high level of nutrition that if she's been kept on a minimum diet.

TEASING SHEEP

The term teasing refers to putting a ewe together with a ram during the breeding season, but preventing the ram from actually getting the ewe pregnant. The idea behind teasing a ewe is that ewes produce fewer ova during their first heat cycle than during subsequent ones. Therefore, it would be nice to bring the ewe into heat by having her with a ram who can't really fertilize her, and then to put her with the ram for the real breeding on her second heat, with the likelihood of multiple ova and multiple births increased.

One way to accomplish this is to have the ram penned next to the ewes so that they can see and smell one another. This will work to some degree, but has the disadvantages that the ram may bust out and breed the ewes on their first heat, and that there is no sure way to find out when any given ewe has come into heat. We tried this one year and didn't like it because it didn't seem to do the job we hoped for.

Another way is to have the ram actually in with the ewes. A big advantage here is that you can outfit the ram with a marking harness. He will leave a colored mark on the ewes back when he mounts her, so you will be able to date her first heat period. This also has the advantage over the "through the fence" method that the ram will actively seek out even the shy ewes who might not visit him through the wire. How does one prevent the ewes from becoming pregnant if the ram is mounting them? There are two ways.

One way is to buy a rig called a Tamm Ram jacket (Sheepman Supply Co., P.O. Box 100, Barboursville, VA 22923, $16.85 plus shipping) and put it on the ram. It is a kind of jockstrap for rams that allows them to urinate freely, but not to impregnate a ewe. It has the disadvantages of lack of cleanliness. You must also have a marking harness on the poor ram so that he ends up looking like a skydiver who lost his parachutes, not to mention the effect of all of the paraphernalia on the ram's libido.

Another way, and the one we prefer, is to have a ram surgically sterilized by a veterinarian. Make sure the veterinarian understands that you want a vasectomy (bilateral vas ligation for you medical term fans) done, not a castration. Our vet charged us $15 to vasectomize a ram for a teaser. We chose a ram who had good enough wool that he pays his feed bill from wool alone so there is no loss from having a nonfertile ram around the place. If you don't feel that you can justify a teaser ram, maybe you'd want to try one of the other methods or share a teaser with neighbors.

Does teasing really work? A veterinarian told us that sheep-

Table 4-4. Lambing Percentages for Unteased and Teased Ewes.

Ewe Group	Lambing Percentage
Unteased ewes	100%
Teased ewes with Finn. blood	215%
Teased ewes without Finn. blood	169%

men believed that it did, but he didn't actually know of any real evidence based on scientific experiments. Was it folklore or reality? The vet thought it sounded reasonable, and so we did our own experiment last breeding season. We had planned to tease all of the ewes and see if our lambing percentage (average number of lambs per ewe as a percentage) was higher than the year before. This was not a well-designed scientific experiment because we really should have had a control group of ewes who were not teased, but were bred on their first heat. For obvious economic reasons, we were reluctant to do this. Our experiment was done correctly for us. One night an unvasectomized ram jumped the fence into the teasing pasture and randomly selected and bred some of the ewes before morning when we kicked him out. We were made at him, but he made us responsible scientists.

We are through lambing for this year and can report that teasing does indeed work, very well as a matter of fact. In Table 4-4 we have three groups, unteased ewes of all breed types, teased with ⅛ to ⅜ Finnsheep ancestry, and teased ewes with no Finnsheep ancestry.

Table 4-4 speaks for itself about teasing's advantages. It is also worth commenting on the effect of breed type. Finnsheep are famous for producing litters of lambs, but the lambs are very small and adult sheep are not an ideal market type. With teasing, a very small portion of Finnsheep ancestry gives a very healthy lambing percentage without some of the disadvantages of pure or half Finn sheep.

A high lambing percentage is not all good. A high birthrate means a bunch of triplets, and it is a rare ewe who can produce enough milk for three. You'll have to bottle feed a bunch of lambs if you get a high lambing rate. If you have the labor or the equipment to handle a lot of orphans, then go ahead and try it. Good luck.

Be sure to have the ram vasectomized quite a while prior to breeding season because live sperm will be retained for a long time. About three months is probably adequate precaution.

MANAGING PREGNANT EWES

A ewe's pregnancy can be roughly divided into two periods—the first 15 weeks and the last six weeks. Each period requires different methods of management and feeding.

The first and longest period requires less management. The goal at this time is to see that the ewe gains a slight amount of weight without becoming fat. As 70 percent of fetus growth is in the last six weeks of pregnancy, a ewe who is allowed to become overweight during the first 15 weeks of pregnancy will be grossly overweight by delivery time.

Excessively heavy ewes will often experience lambing and milking problems, and trying to put a pregnant ewe on a reducing diet is to invite the death of mother and/or lambs. Therefore, it is important that the shepherd carefully evaluate the condition of his ewes even during the first 15 weeks.

If ewes are in medium to heavy fleece, it is often not enough to judge their weight gains or losses by looking at them. Getting in the habit of feeling the backbone and ribs gently with the fingertips by pushing down through the wool to the skin is a good practice. Once a person becomes accustomed to how the ewes feel in general, any weight loss or gain can be detected in this manner.

According to research statistics, a ewe should be maintained at a nine percent protein level in late pregnancy, regardless of the components of the ration. In general, this can be achieved through feeding good alfalfa hay, or a lesser quality hay and a little grain.

If the shepherd finds a big difference in ewe weight and/or age within the flock (such as with yearlings and older does), it is a good practice to separate them into different feeding groups. This ensures that yearlings, who need extra feed because they are still growing as well as maintaining their lambs, and older or shy ewes, who eat more slowly or less aggressively, will be able to get a proper amount of feed.

Regardless of age, researchers have found that ewes suffer no ill effects from being fed every two or three days, as long as protein levels and weight are maintained. During late pregnancy, daily or twice daily feeding is recommended. It is also extremely important that pregnant ewes have access to all the salt and minerals they need, as the fetuses are drawing on the ewe's reserves and increasing her requirements.

In addition, exercise in moderate amounts is necessary to maintain the ewe's muscle tone and to aid in preventing the ewe from becoming overweight. Even in a medium-sized lot, where ewes may be housed for the winter, exercise can be encouraged by placing the feed bunks some distance from the shelter to necessitate walking. Another trick is to sprinkle a little shell corn at a distance for the ewes to pick up in their leisure time.

Given the proper level of feed and exercise, it is unlikely that ewes will experience any health problems. The shepherd must be alert throughout pregnancy for first signs of illness or going off feed. Any such condition should receive immediate attention. Illness coupled with the heavy drains on the body's resources due to pregnancy can quickly deplete a ewe's health, exposing both mother and lambs to danger. If worming is necessary, be sure to choose a product approved safe for pregnant ewes.

As the first ewes come within six weeks of lambing, they should be separated into a group of their own. As stated before, almost three-fourths of the fetus's growth occurs during this period. This necessitates stepping up the ewe's nutritional level. It is especially important that the ewe now be put on a feeding level that will cause slow but sure weight gain. In many instances, this is done by beginning with ½-¾ pounds of grain per ewe per day, slowly increasing the amount to 1-2 pounds per ewe per day.

If the ewe is expected to have a single lamb, is small, or is overweight, maintaining her at 1 pound per day is usually sufficient. If the opposite is true, and if she is a yearling, maintaining her at 1½-2 pounds per day is best, whichever suits her body frame.

Unless the sheep are maintained in an environment that allows for total shearing, ewes should be crotched and tagged three to four

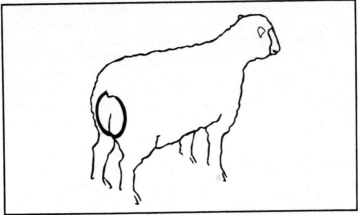

Fig. 4-15. The area to be tagged prior to lambing is outlined (drawing by Margot Mayr).

Fig. 4-16. The area to be crotched (drawing by Margot Mayr).

weeks before they are due. Tagging frees a ewe's rear from long, dirty wool and aids the shepherd in judging just when the lamb will be due (Fig. 4-15). Having the rear tagged also makes it easier and cleaner to assist a ewe in lambing, should she need it.

Crotching, or shearing the wool from around the udder, makes it easier for the lambs to find the teat, without latching onto a lock of dirty wool (Fig. 4-16). In addition, the shepherd can more easily keep an eye on the condition of the udder.

If a ewe has a very wooly face, carefully shear the wool away from her eyes. This helps a ewe to be a better mother as she can count her lambs more easily and keep better track of their whereabouts.

It is very important that shearing, or virtually any handling at this stage, be done very gently and with as little alarm to the ewes as possible. Otherwise, abortion may result.

About the same time that the shearing is done, the building should be prepared. In general, this consists of completely cleaning and disinfecting all sheep quarters, with special attention to those to be used for the lambing pens, closest ewes and very young lambs. Setting up lambing pens and heat lamps early can save a lot of time and trouble at critical moments later on.

When the closest ewes are due within a week, it is important that an effort be made to keep a close eye on them. It may even be necessary to pen them inside at night or if they will be left unattended for much of the day. In this way, no lambs that arrive early or at odd hours will be dropped in a snowbank or lost in a dark corner.

After all this labor and patience, the shepherd has only to wait a matter of a few days until the first rewards of his labor can be counted and tagged.

LAMBING

If you have accurate breeding charts, you can expect ewes to lamb 147 to 153 days after service, varying with the individuals age, weight, and number of fetuses. A ewe who is very close will exhibit a slack, loose vulva (often with a thick mucous discharge), a full, tight udder (except in the case of first-time lambers, who often have only small udders), and a sinking in of the sides both in the flanks and on both sides of the tail head.

If you firmly suspect that the ewe is ready to lamb, cut out her grain feeding for the day, as it may lead to constipation or udder congestion. Many shepherds also discontinued feeding grain, but not hay, for the first 24 hours after birth for the same reasons.

Check the ewe for milk by grasping the teat about midway up and gently squeezing several times. If milk appears, and it should if she's ready to lamb, milk a few squirts out of each teat to remove any germs that may have entered the milk through the teat canal. You don't want the lamb's first drink to be unhealthy.

At this time, be sure the ewe is enclosed in a light, draft-free area of the barn with no access to the outside if the weather is bad. She should be checked a minimum of every four hours until labor begins.

Ewes can and do lamb at any time of the day or night, although many find that the greatest activity tends to be before midnight and after 5 or 6 A.M.. Unfortunately, there's no guarantee.

When actual labor begins, the ewe will become restless, getting up and down frequently, pawing the ground, and perhaps looking and calling for a lamb. It is important now that she be checked very frequently, but not shut into a lambing pen, as she needs plenty of room to deliver once she lies down. Confining her in a pen will allow the fetal fluids to soak the pen, chilling the lamb.

In a few cases, however, when the ewe is abnormally wild, it may be best to pen her in case she needs assistance at a critical point later. Mother and lamb(s) can always be switched to a dry pen once the lambs have been safely delivered.

If more than one ewe is close, especially if both have had lambs before, the shepherd must check to be sure one doesn't steal the other's lamb. Older ewes, "knowing" they will lamb soon, may gladly accept another ewe's lamb, even giving it their milk.

A ewe may continue her restless pacing and pawing for several hours. It depends on the individual, but eventually she will get down to business.

The ewe normally will lie down (a few will deliver standing) and begin to strain with increasing frequency and vigor. At this point, some ewes object to having a human audience and will try to suppress labor until the shepherd leaves, conceals himself, or is perfectly still. The ewe will also try to isolate herself from the other animals as much as possible. Actual labor may take from one to two hours, with some ewes delivering in a matter of minutes.

The allantoic sac, the first water bag, slowly appears. This clear, fluid-filled bag should not be broken by the shepherd under any circumstances. Once it is broken, the lamb tries to breathe and will drown if he is not quickly delivered.

The water bag may hang from the ewe's vulva for quite some time or it may quickly break. In either case, the second water bag, the dark-colored amniotic sac, appears next. It, too, may burst or remain dangling.

There is no reason for concern if a lamb appears before the second water bag. Either this bag or another lamb will follow him.

Once the birth canal has been dilated and lubricated by the water bags, it is prepared for the passage of the lamb(s). The ewe normally expels the lamb's forelegs and head first, with the lamb's muzzle lying neatly on its forelegs. Once the shoulders have been delivered, the rest of the body should slide out easily.

The ewe should then get up and clean the mucus from the lamb's body, especially his nose, thus stimulating him to breathe. If she fails to do this, it should be quickly done by the shepherd.

If the cleanings (the fetal attachments to the uterine wall) do not appear soon, the ewe may begin to strain again and present another lamb. After the last lamb is delivered, the cleanings should appear. These look like a collection of spongy round "buttons" bonded together by a network of membrane. The ewe should be allowed to expel the cleanings herself. They should then be removed from the lambing barn and permanently disposed of as they sometimes harbor disease.

Once all the lambs have been dropped, put the family in a well-bedded lambing pen equipped with a heat lamp, if it is very cold, to help dry the lambs. Unless it is bitterly cold, remove the lamp when the lambs are dry, as too much reliance on artificial heat will breed low resistance to chills and pneumonia. At this time, offer the ewe a bucket of warm water to drink.

The lamb's umbilical cord should be thoroughly doused in iodine to prevent infection, and he should be given any injections for vitamins, scours, overeating, etc., that have been agreed upon earlier with the vet. Mark the lamb(s) with an ear tag so you know who belongs to whom.

A good mother will help the lambs learn to nurse by arching her back, pointing her teats forward, and gently pushing the lambs toward them. Once the lambs get a taste of colostrum, they're on their way.

Leave the ewe some water in a bucket high enough to keep lambs from falling in. Then let the family get acquainted. They should remain in the lambing pen for at least two days to allow the shepherd to keep an eye on the health of both lamb(s) and ewe as they gain strength and learn to recognize one another.

As every lambing presents a number of different events, it is important to realize that not every birth follows the above pattern. Until the shepherd gains experience with normal lambings, it is difficult to judge when an ewe may or may not be in trouble. The key point is patience. More often than not, nature will take care of the birth process with little outside interference.

A few potentially troublesome situations should not be overlooked. These can be identified and rectified even by someone relatively inexperienced.

If the first and/or second bags appear and break and the ewe quits laboring entirely or strains for 15 to 30 minutes without apparent effect, the shepherd would be justified in checking out the situation. Some research indicates that a lamb cannot live in the ewe for longer than 15 minutes after the first water bag has been broken, so close attention is required. At times, however, even lambs who have been jockeyed around for over an hour while still attached by the umbilical cord, are fine once delivered.

It is often just as well in this situation to go ahead and locate the problem rather than wait and wonder. Even if you intend to call a vet, it helps him to know the situation.

If the ewe is standing, get someone to hold her head and press her body gently against a wall or lambing panel so she can't squirm and hurt herself as you work. If no help is available, tie her with a halter securely to one corner of a lambing pen and "squeeze" her in with another panel.

At this stage, strength will count little. It is often best to have someone with small hands check a ewe. Be sure the examiner wears gloves. If he or she has a deep cut on their hand, an infection from a ewe may enter the cut.

Thoroughly wash your hands and arms (be sure your fingernails are short) and leave some gentle soap and water on one or coat it with Vaseline petroleum jelly or a similar safe lubricant. Be sure to remove all jewelry.

Kneel behind the ewe and with your fingertips first, gently, and slowly push your hand into the ewe's vagina. If she is straining, stop. Only push as she rests, not when she strains.

Often the lamb is right there, and it takes only a few seconds to tell what the problem is. If the nose is there, resting on both forelegs, try to gently ease him out by grasping both forelegs and pulling out and slightly downward in the direction of the hocks.

If the lamb is not presented normally, and you don't want to call a vet until you're sure you need him, very gently move your hand along the lamb's body. A ewe's reproductive tract is extremely fragile. Some vets compare it to tissue paper, and sudden, bungling movements can tear through the wall. Proceed slowly and with much care while the ewe is resting.

If you feel the tail and hind legs of a breech birth, pull the lamb out as quickly as possible. Such presentations, if delayed, often result in drowning. Again, don't be rough, as the lamb's rib cage sometimes sticks on the ewe's pelvis. The lamb must be partially rotated to be delivered.

If the lamb's head is tipped back or one or both legs are curled back or under, very gently push the entire lamb back toward the uterus, where you will have a bit more room to work. Slowly bring the head down or the leg(s) around to the proper position. Make sure that both are front legs. You can tell by the shape the way the joint bends. Also, be sure they belong to the same lamb. If the lamb is too slippery, it may be necessary to loop a clean, disinfected piece of wire, nylon rope, or small chain over the head or legs.

Once in the proper position, the ewe should be able to deliver the lamb, unless she is too tired. In this instance, you can help by keeping one hand on the rope or string outside and one hand inside on the lamb's legs. Gently exert pressure out and down only when the ewe strains.

If the ewe is no longer straining, an injection of the hormone oxytocin, at the recommended dosage, will cause labor to begin again. Oxytocin is also useful in the case of a ewe who has stopped labor completely after passing one or both water bags, and no lambs are within reach.

If a lamb with one foreleg back cannot be pushed back into the uterus, it is often a good idea to try to deliver him as is. Begin with gentle, steady pulls as the ewe strains and increase to pull as the lamb moves, remembering the fragility of the ewe. If the lamb won't move even after some time, the reproductive tract may be too dry. If more lubrication doesn't solve the problem, it is a matter for the vet, unless you have quite a bit of experience. Brute strength will do no good.

If the ewe expels a lamb's head and one foreleg and the lamb is stuck, try lubricating the head and gently forcing the lamb back through the vagina into the uterus where you will have more working room. Again, only push when the ewe rests.

If the head is too swollen, and this can happen in a matter of minutes, lubricate the head and try to deliver, pulling on the extended foreleg. If the head is swollen and both legs are back, it will be next to impossible to deliver the lamb alive unless it can be returned to the uterus and straightened out. Trying to pull it in such a position will do nothing but harm the ewe. A ewe who has this or any lambing problem can be assumed to be carrying more than one lamb and should be checked.

In addition, any ewe who has had a difficult delivery, requiring human intervention, is bound to have increased chances of infection in the reproductive tract. She should, therefore, be given antibiotics for at least two days in the form of injections and/or uterine boluses.

POST-LAMBING DIFFICULTIES

A ewe may experience post-lambing difficulties that will affect her mothering ability. A ewe who has had a difficult delivery may feel indifferent toward her lambs and not allow them to nurse. The same is true if her udder is badly congested or swollen so that it is painful for the lambs to butt it and nurse.

Some ewes, especially first-time lambers, just don't seem to want the lambs and need extra attention. The shepherd will have to hold the ewe and help the lambs nurse until she gets used to the idea, giving them extra time in the lambing pen. With very stubborn ewes, it is necessary to immobilize their back legs with hobbles or to put them in special stanchions to keep the ewe from kicking the lambs away from the teat. At times, setting the ewe up on her rear, as in shearing, and putting the lambs on the teat is the only way to get them going. In the case of the congested udder, the ewe should be receiving medication, too.

If the udder seems far too pendulous, is hot, changes color, looking badly inflamed or even blueish, or shows ropy or bloody milk, immediately get antibiotics from the vet to reduce this mastitis. In the meantime, lambs may have to be fed artificially or even grafted onto another ewe. It is easier for the future, however, if the lamb(s) can stay with the ewe and be supplemented on milk replacer. That way, the lamb's constant massage of the udder aids in reducing congestion, and the ewe doesn't go without a lamb to raise altogether.

If a ewe has big twins or more, it may be good to remove all but one. Be sure to mark for culling, as mastitis not only usually returns, but may spread if caused by bacteria. It is, therefore, necessary that the ewe and her lamb(s) be kept from the other ewes. Lambs may spread the organism if they try to nurse other ewes, as sometimes happens.

If no mastitis is present, but the teats are too big for the lamb's mouth, they should be milked down as often as necessary. The milk is saved in ice cube trays for later arrivals who may need it.

If a ewe has very little milk, a significant increase in the amount of feed, especially grain, may be helpful. At times, ewes are slow in coming to milk and may show a marked increase after the first 24 hours. If this does not happen, oxytocin injections as prescribed by the vet can increase milk letdown. Lambs may have to be supplemented with another ewe's milk or with artificial lamb milk replacer in a bottle to prevent starvation.

In cases of delivery involving much pulling, a ewe sometimes prolapses, or expels her uterus, vagina or rectum, often with little warning. If you lack experience in such cases, call the vet at once and try to keep the area clean. He will put the organ back into its proper place and sew up the ewe. Such an animal is a prime candidate for culling as prolapse may reoccur and can be inherited.

Lambs may also have problems the first few days of life, chief among them starvation. The most reliable method of judging whether or not a lamb is hungry is to put your finger in its mouth. If the mouth is warm, the lamb is nursing. If not, track down the problem without delay.

A healthy lamb will also look plump and not bleat excessively. He will not appear weak or sleepy all the time.

In the case of an orphan lamb and/or a lamb that is born too weak to nurse, the following method may save the lamb's life.

Get a plastic syringe and a length of rubber tubing about a foot long of the type used in human catheters. Place the lamb on his stomach in a level position with his head and neck stretched out.

Begin working the tubing into his mouth, from the front if possible, pointing the end down and keeping the body, head, and neck straight. If the tubing is headed into the stomach as it should be, there will be little resistance. If, however, it heads into the lungs, the tubing will hit a snag and begin to back up, in which case you must withdraw it and start again.

Once the tubing has disappeared easily almost up to the syringe, fill the syringe with warm (not hot) colostrum and very slowly replace the plunger, forcing colostrum into the lamb's stomach. Usually three or four 12-cc syringefuls of colostrum at four-hour intervals will quickly bring a change in the lamb's condition. Some shepherds also use a dextrose injection on weak lambs.

Ideally, colostrum from the mother, or at least that of another ewe, should be used. If there is none available, however, try feeding a mixture of whole milk, egg, sugar (with a drop of brandy for a really weak lamb), and ¼ teaspoon of mineral oil as a laxative. The other ingredients should be in amounts equal to the egg. Cow colostrum is good if you can get it.

Injections of vitamins A, D, and E, combiotic, iron, and possibly antienterotoxemia (overeating disease), will help make up some of the protection a lamb normally gets in colostrum. Without at least one feeding of colostrum, however, a lamb's survival chances are small.

As soon as the lamb gains strength and can stand, switch him to a bottle. He should have colostrum for at least 24 hours.

If a lamb seems hungry and the ewe's milking ability or disposition is not at fault, the lamb may just not know how to nurse. He may be going to the udder, butting it and making noises while wiggling his tail madly and never have the teat at all.

Stand him up near the udder and tickle under his tail. When he begins to make sucking noises, run your fingers to his mouth and guide him to the teat. Sometimes it is necessary to repeat this procedure every few hours for a day or two to keep lambs from starving.

If there are rejection or milking problems after the lamb has gained some strength, you can't induce the ewe to take the lamb, find it a foster mother, or feed it artificially. If the ewe continually tries to reject the lamb, introducing a dog into the barn near the lambing pen often raises a ewe's protective instincts. If this fails, the ewe can be immobilized until she accepts the situation. Remember to supplement the lamb with milk replacer or with another ewe's milk in a bottle if he cannot nurse.

If the ewe continues to reject the lamb even after several days, it is often less time consuming to find him a foster mother than to feed him yourself. If you have a ewe who has very recently lambed and who has plenty of milk and only one lamb, or if she has lost a lamb, she is a prime candidate.

The most reliable method, if the ewe's lamb is dead, is to skin her lamb and tie the fleece to the orphan. That way, the lamb smells "right" to the mother.

In the event that the ewe has just lambed with a dead lamb, the orphan should be totally immersed (except the eyes, nose and mouth) in warm water and rubbed immediately with the birth fluids of the dead lamb. If necessary, his legs can be loosely tied for a half hour or so so that the ewe doesn't become suspicious as to why the newborn gets the hang of walking so quickly.

If you don't have a ewe in this position, try rubbing the ewe's nose and the lamb's body with something strong-smelling like perfume or deodorant.

The ewe must be confined until she accepts the situation. A dog can be brought close to the pair to raise the ewe's maternal, protective instincts.

If the lamb must be raised on artificial milk entirely, lamb milk replacers are available at most large feed stores. Calf replacers should be avoided as they are usually deficient in fat.

It is not necessary to feed milk replacer warm. Feeding the recommended amounts cold will retard bacterial growth in the milk and keep the lamb from consuming too much too quickly and possibly contracting overeating disease.

Orphan lambs should be started on creep feed as a supplement as soon after birth as they will accept it. Keep orphans in separate groups with others of the same size. House them in a dry, warm area.

Watch for scours. This condition should be treated without delay as death is usually rapid.

A lamb's tail may become "pinned" to his body with the discharge of his first sticky manure. If the ewe doesn't clean it away, the shepherd has to. Often a coating of Vaseline helps soothe raw tissues, and a light sprinkling of dirt or sand will prevent a recurrence.

Lamb's eyes should be checked for signs of entropion — inverted eyelids. In these cases, the lashes turn inwards, brushing and constantly irritating the eye. If left untreated, blindness results.

Many cases of this inherited trait can be cured by simply

remembering to pull the eyelid away from the eye several times a day and treating it with some antiseptic powder. Nonresponsive cases should be surgically altered in a simple procedure you can learn from another shepherd or a vet.

Lambing pens should be disinfected thoroughly and ideally allowed to rest between ewes. Any ewes that abort should be kept strictly confined from pregnant ewes and the aborted tissues burned or buried.

When problems, if any, have been resolved, the ewes can be turned into groups of 10 with lambs of the same approximate size. Although they should still have a dry shelter free of drafts, young lambs can be allowed access to the outside except on the worst days.

Ewes with twins should be getting at least 2 pounds of grain per day and all the alfalfa hay they want. Ewes with one lamb need only 1 pound of grain and approximately 4 to 5 pounds of alfalfa. The amount will vary with the individual's weight, milking ability, and the number of lambs. Even though they are out of the lambing pens, be sure to watch both ewes and lambs for possible problems.

TENDER LOVING CARE AND MILK REPLACER WORK WONDERS

Nothing is as cute as a baby lamb, and nothing is more heartbreaking than a dead one. Whether you are raising sheep for profit, your deep freeze, or the kids, you still run up against the problem of orphan or rejected lambs. Don't try to find an answer to the rejected lamb. No one seems to know the reason. Many have theories, but no one has the answer.

Tender loving care and a good milk replacer works wonders. We'd like to be able to assure you that you will save them all with this method. From experience, we know it isn't so, but it will certainly raise your percentage of survival.

The most important lesson in bottle feeding a lamb is don't overfeed it. They try to make you think they have the appetite of a horse. Really they don't. Overfeeding has killed more lambs and caused more tears than we care to remember. They'll cry every time they see you or hear you. This is really a cry for company. They'll keep eating as long as you'll hold the bottle.

There are many formulas for lamb bottles. Over the years we have found Felco Land O'Lakes lamb milk replacer or Alberts Lama work well. Maybe others just as good are on the market.

Check with your local feed companies for customers who order lamb milk replacer. Try to do this before your lambing begins. The expense of a 25-pound sack is high. Try to make arrangements with one of them to buy a pound or so, if you should need it.

Milk replacer is a dry milk. If you haven't a blender borrow one. It is a must. The usual mixture is 1 cup of dry milk to 4 cups warm water. Feed this at about 40 degrees Fahrenheit.

We use an animal nipple on a pop bottle. The nipples aren't that expensive; get one before you need it.

When feeding a lamb, keep its head up by holding the bottle above its head, at about a 45-degree angle. Some take to the bottle naturally. For those that don't, put them between your legs. That way one hand can hold the bottle. The other can force its mouth open and insert the nipple. This also works well for the weak lamb. It forces the lamb to stand while eating. This strengthens its legs and gets it moving faster.

There are two methods we use to get a chilled or sickly lamb to eat. The scientific way is a ball syringe with a long plastic tube. This is filled with warm milk and guided down the lamb's throat to the stomach.

Give it a teaspoon of whiskey. This starts the throat moving and warms it. Place the lamb between your knees and insert nipple. Be sure to keep that head up.

If a lamb dies or is rejected, milk out the ewe the first day. This is the best milk for the newborn that has to be bottle-fed. We freeze it in an ice cube tray. Each cube is about an ounce. When needed, melt the cube slowly to keep the colostrum from thickening.

Use tender loving care. Talk to the lamb while feeding it. Its mother does. It needs attention. Before you know it, the sound of your voice will bring it running, ready to eat.

For a day-old lamb, 2 ounces is enough every hour and a half. Gradually increase this as it gets older. Remember, better too little than too much. Don't worry; it can go through the night with a late and early feeding. As you increase the amount, lengthen the time between feedings until it is eating about every three or four hours by the third or fourth day. After a couple of weeks, it should be getting between eight and 12 ounces, depending on size, every six hours. Your lamb should be ready to wean in 30 to 35 days or when it weighs 25 to 30 pounds.

Expose your lamb to lamb pellets or grain and hay within the first week. Sprinkle this with a small amount of dry milk replacer. Curiosity will make it try it. The sweetening in the replacer makes them want to eat it.

Nature takes care of its own. Don't keep your lamb in a warm place, unless it has been chilled. If so, after it is on its feet, move it to cooler temperatures, but out of the draft. Back to the barn, with a heat lamp if necessary, is best. Place them near the ewes, but in separate pens. Nursing ewes do not like freeloaders.

Tender loving care takes time. You won't save them all, but those you do are well worth your effort.

DOCKING AND CASTRATING

Right now, with little lambs on the ground, is the best time to insure additional health and profits for your flock through the simple procedures of docking and castration. Even if you may be unaccustomed to these practices, there is nothing complicated or mysterious about either of them. Both are basic management techniques that every sheep raiser should be familiar with.

Since undomesticated sheep have gotten along with these extra appendages for ages, you might well wonder why they deserve any attention now. The main reason is health. A lamb with a tail is 100 percent more likely to have manure buildup between his body and tail, not only creating an extremely wet, dirty breeding ground for disease, but leaving the lamb highly susceptible to fly-strike, maggots, or screwworms, any of which mean an extremely serious problem, if not death. In addition, a long tail hampers the ram's breeding efforts and makes it more difficult to assist a ewe at lambing.

With castrating, the issue is less clearcut. If the lamb is to be sold to a packer or if he will not be slaughtered before five, or at the latest six months of age, you will most likely want to castrate. Buyers generally knock $1 off the bid price for ram-lambs. Those who slaughter at home after the ram is five or six months old may find that the meat has a distinctly strong taste, caused when testosterone finds its way into the meat—the same reason buyers dock their price.

On the other hand, wethers (castrated rams) will grow a bit more slowly than rams. Wethers have more weight in the desirable cuts of the hindquarters as contrasted to the ram.

Once you have made the decision, try to do both operations before the lamb is over 10 days of age. Be sure that the lamb is healthy before attempting either of these procedures as it may be too much of a shock to a weak lamb.

Have a helper hold the lamb by securely grasping a fore and hind leg in each hand and set the lamb's rump on a table or support his weight in some way with your body. In the case of a ram lamb,

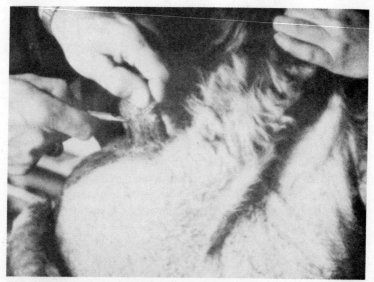

Fig. 4-17. Cutting away the bottom of the scrotum.

castrate first and dock second. Otherwise, the lamb may put too much stress on the docking wound and cause bleeding.

Being careful to miss the testicles, simply cut away the bottom one-third of the scrotum with a disinfected (five percent carbolic acid solution or coal-tar dip used according to directions on the package), sharp knife or scalpel (Fig. 4-17). Expose the testicles by gentle pressure at the base of the scrotum where they join the body cavity (Fig. 4-18). Keeping one hand there to prevent either testicle

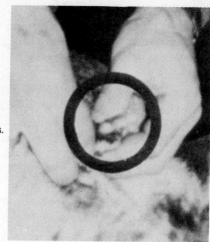

Fig. 4-18. Exposing the testicles.

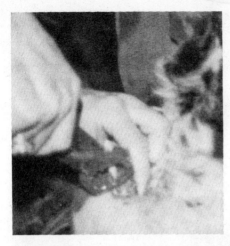

Fig. 4-19. Pulling the testicles.

from slipping back, grasp the exposed end of a testicle and pull it free of the body (Fig. 4-19). Repeat the procedure with the other.

The testicle can be pulled with an instrument designed specifically for that purpose (a three-in-one), with your fingers, a small pair of pliers, or any instrument that will not cut through the tissue. Whatever you choose, be sure that it is thoroughly disinfected before use.

Thoroughly cover the wound with a prepared disinfectant such as iodine. If the procedure is done during fly season, be sure to also

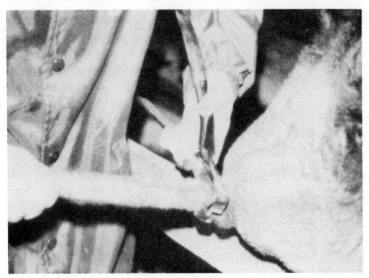

Fig. 4-20. Placing the emasculator.

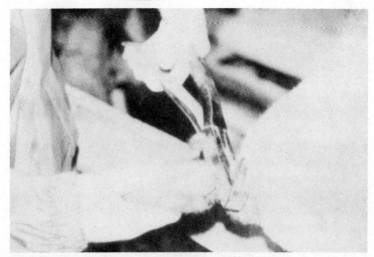

Fig. 4-21. Squeezing the emasculator until the tail comes off.

apply a repellent each day until the area is completely healed.

For docking, there are a wide variety of instruments available, ranging from the knife or hatchet to the bloodless elastrator that uses rubber bands. In Figs. 4-20 and 4-21 an emasculator was used.

In this simple procedure, the instrument is placed on the tail, curved end down, and in such a position as to give the lamb no more than a 1-inch tail (Fig. 4-20). Steady pressure is applied until the tail drops off (Fig. 4-21). The emasculator is left on an additional 30

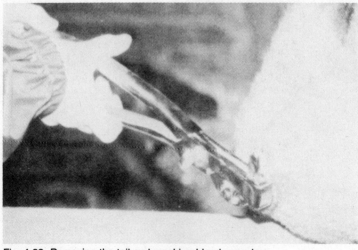

Fig. 4-22. Removing the tail and crushing blood vessels.

seconds to be certain all blood vessels are crushed and little or no bleeding will occur (Fig. 4-22). Disinfect and apply a fly repellent liberally.

That's all there is to it. Just be sure not to put these lambs in with rougher, bigger ones until all danger of bleeding has stopped. Any serious bleeding, a rare occurrence with the emasculator, can be stemmed by applying a pressure bandage or tying a string above the wound for a few minutes. Watch the area closely for several days for signs of infection or fly-strike. A few minutes spent on these easy procedures now will later pay for themselves many times in healthy, tasty lambs.

FEEDING OUT LAMBS

Once lambs are a few days old, they begin to explore the world around them, and one of their first interests is what the ewe finds so absorbing in that grain pail and hay rack. Take advantage of the young lamb's inquisitiveness by setting up a special area of the barn for lambs only: an area that is well-lighted, warm, comfortable, and full of fresh, palatable feed at all times.

Because a lamb makes his most efficient gains the first 100 to 120 days of life, creeping feeding will usually pay off better than putting on lamb weight through ewe milk alone. If lambs are born late enough to be able to take advantage of lush pasture, however, they may make cheaper gains on that feed source.

If a creep is to be used, it should be located in an area of the barn where ewes naturally collect, such as near the feeders or water tanks. By the time the lamb is a week old, he will be getting bold enough to leave his mother for short periods of time to go investigate the creep, especially if he sees other lambs using it. Feeders in the creep should be lamb-sized, but constructed so that the lambs cannot stand in their feed. Often a horizontal bar across the top of the feeder will keep lambs out as well as is possible. The USDA, county extension office, and many state colleges can usually supply any number of plans (Fig. 4-23).

As far as a palatable, properly balanced ratio is concerned, the producer may invest in a complete feed or mix his own. Complete lamb feeds are usually pelleted and balanced at about 13 or 14 percent protein. There may be some problem in getting lambs to accept the complete feed initially, a problem which can be quickly overcome by mixing in a small amount of cracked corn, rolled oats, granulated molasses, or a similarly palatable treat. If the producer has the equipment and time and can mix his own lamb ration more

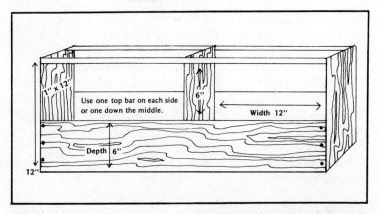

Fig. 4-23. This durable creep feeder is easy to build. As lambs grow, it can be elevated on legs to grow with them. According to most specifications, allow one linear foot for each lamb (drawing by Margot Mayr).

cheaply than he can buy it, the correct balance is simple to calculate.

To figure the main elements of protein and TDN (total digestible nutrients), one has to have a copy of the nutrient requirements of sheep. A copy can be obtained from the National Research Council, whose figures are used here. The address is: Agricultural Board, Committee on Animal Nutrition, Subcommittee on Sheep Nutrition, Washington, DC. Figures are also available in many feed booklets or magazines.

As an example, let's say the producer wants to formulate a ration for an 80-pound feeder lamb using homegrown grain and hay. He sees that this lamb needs:

Feed pounds/day	TDN	Crude protein
3.4	2.1	.36

He has decided to use late-bloom alfalfa hay and Number two shell corn in his ration. According to the tables, these contain:

	TDN %	Crude Protein %
Corn	77.4	8.9
Alfalfa	48	14

To achieve the necessary 3.4 pounds of feed/head/day, the producer decides to feed 1.8 pounds of corn and 1.6 pounds of alfalfa. To see if this ration will meet the requirements of his lambs he figures:

	TDN	Protein

Corn:
$.774 \times 1.8 = 1.39$ pounds $.089 \times 1.8 = .161$ pounds
Alfalfa:
$.48 \times 1.6 = .768$ pounds $.14 \times 1.6 = .224$ pounds
            Total TDN   Total Protein
$1.39 + .768 = 2.158$ pounds $.161 + .224 = .385$ pounds

Referring to the beginning of the problem, we see that this ratio will meet specifications for both TDN and protein. It will supply .025 pounds more crude protein than is required. Whether a complete or home-mixed ration is used, high quality roughage, preferably in the form of alfalfa hay, is important to stimulate development of the rumen and should be included in every creep ration.

Lambs should be fed only as much concentrate and roughage as they will clean up in one feeding. Any left over should be fed to the ewes while lambs get the freshest feed possible. Salt and mineral must also be fed the lambs free choice at all times, along with access to plenty of clean water.

If lambs have not been vaccinated for overeating disease previously, this may be a good opportunity. For the purpose of keeping records on which ewes, lambs, and ratios are the most profitable, the producer should continue to weigh the lambs at specific intervals every week or two after birth through marketing or first lambing.

No one can say specifically what health problems a lamb may encounter between birth and market. Pneumonia and overeating disease are often major problems with listeriosis, urinary calculi and rectal prolapse being lurking possibilities in certain flocks.

Hand-feeding lambs once or twice daily gives the producer an excellent opportunity to keep an eye on his lambs and to catch any healthy problems before they become very serious. Before a healthy male lamb is 10 days or at least two weeks old, the decision of whether or not to castrate should be made. All lambs should be docked by two weeks of age.

By about three months, the producer should wean the lambs. Separate ewes from the sight, smell, and sound of the lambs completely. It is generally easier for the lambs to adjust to this stress if they are left in familiar surroundings and the ewes moved. At this time, the producer must decide which ewe lambs he intends to keep as replacements for his flock and which to sell.

Usually the earliest born, healthiest twins, replacement ewe lambs should be separated into a group of their own and receive a ration that will allow them to grow to their full potential without laying on harmful fat. They may receive the same ration as the feeder lambs, but in lesser amounts along with all the high quality alfalfa they can eat.

The producer must remember to change lamb rations only very slowly. Gradually mix a new ration in with the old until the change is complete in about a week or 10 days. Doing so will prevent stress and related health problems.

By weaning or shortly after, feeder lambs should be making their best gains on concentrates and will eat correspondingly less roughage. As when they were creep-fed, be sure feeder lambs get all the feed they want to take advantage of their gaining potential.

If the weather is hot, it usually helps lambs to gain and grow if they are sheared completely at weaning. Certainly it is a good idea to "tag" them or shear around their rear end to prevent manure build up and fly-strike. Shearing face wool helps to prevent eye problems.

If pushed on grain, lambs should be ready to sell by 100 to 130 days of age, depending on their breed and body structure. If the producer has been weighing them at designated intervals, he will note a drop in feed efficiency and rate of gain after this time. In addition, he will be able to feel a definite increase in backfat when he puts his fingertips through the wool and rubs lightly back and forth on the spine. It is generally uneconomical to continue feeding lambs past this date.

MAINTAINING EWES

Early in the lamb's life, the ewe is called upon for greatest milk production, which peaks at an average of 4 pounds per day when lambs are about three to four weeks old. At this stage, the producer must be especially careful to see that ewes are getting enough grain to maintain themselves and their milk production.

As they find the creep more and more attractive, lambs will begin to rely less upon the ewe and neither will be as frantic to keep in constant touch with the other. Even though the lambs may be getting less milk as they grow older, the demands of their increased size upon the ewe are heavy. Her level of feed must be maintained to keep her healthy and to prepare her for breeding back. To this end, just after lambing, she should receive, along with grain, the best quality hay, with decreasing qualities fed only later on.

Near weaning time, ewes may begin to gain weight. They will

be annoyed by the lamb's demands for milk. At this juncture, the producer should start preparing to "dry off" the ewes. To reduce the amount of milk the ewes will produce while avoiding problems with the udder, begin reducing the amount of grain about a week before weaning lambs so that ewes receive no grain four or five days before. Also, cut the hay feeding down to only about ½ pound/head/day at about three days before weaning, switching to low quality, stemmy hay. Likewise, reduce water intake drastically, allowing for very hot weather. Any ewes who are sick must not be dried off as such treatment would harm them.

Some producers prefer to shut off all feed and water from ewes for two or three days prior to weaning. Combined with the lambs' nursing, the udder will dry up well, but often the ewe loses a great deal of condition and may become sick. If the flock as a whole consists of extremely heavy milkers, this may be the only way to avoid post-weaning problems.

Once ewes are separated from their lambs, they should not be allowed to see each other again until both have forgotten—at least two weeks. The only exception would be the development of too much milk in a ewe's udder, in which case keeping her on stringently reduced feed and allowing the lamb in to suck once or twice should do the trick.

Keep an eye on the ewes' udders to see that they do shrink steadily. If a ewe in good health retains a tight, bulging udder, cut her feed and water for two to three days and inject antibiotics daily. If the problem remains or others develop, consult the vet.

After weaning and until they are flushed and bred again, ewes can be maintained on pasture or roughage. Juggle the amount of roughage (with access to plenty of water, salt, and mineral) until you are satisfied that ewes are not gaining or losing too much weight. It is not only permissible, but a good idea, for ewes in good to heavy condition to lose weight during this period, so they can be allowed to gain again when they are bred with no danger of complications from overweight.

Both ewes and lambs must be watched for the unthriftiness that signifies worms, as an individual may become heavily infested at any time. To prevent parasite buildup, it is best to worm lambs at least at weaning and once again before marketing. Ewes should be wormed both during lactation and the dry period before breeding. The dry interval provides a good opportunity to use the wormers specifically forbidden during pregnancy but necessary to insure a complete parasite control program.

SHEARING

A small flock of sheep fits in well with many homesteading and farming operations. One of the drawbacks, from many beginners' viewpoints, is the annual problem of getting the wool off them. Actually this needn't be much of an obstacle, even if no professional shearer is available. Anyone in good health can learn to shear well enough to do his or her own animals.

A cord-type machine, with motor in handle, is usually used on small flocks. Be sure to have enough sharp blades. Dull ones don't cut cleanly and make the job much more difficult. The sheep should be dry and as free as possible of chaff, dirt, and other contaminants. Don't wash them, though, as this removes the natural grease and lowers fleece weight considerably. (The selling price for raw wool is based on grease weight.)

Fig. 4-24. Set the sheep up on its rump.

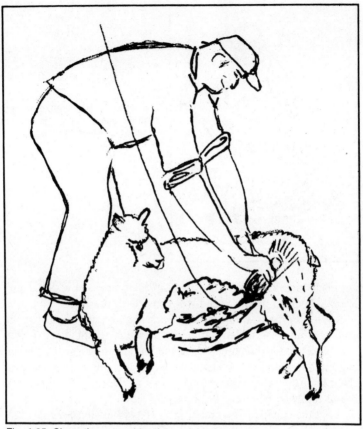

Fig. 4-25. Shear the outer side of the left hind leg.

Shearing should be done on a clean, level floor. An area 6 feet square is large enough. Old discarded rugs are ideal on which to shear. They don't get rumpled as easily as canvas.

To keep the electric cord from getting wrapped around legs, etc., fasten it overhead to your right. With the machine lying on the floor, most of the slack should be removed. A good way to do this, and still get a little more reach, is to tie a long rubber band to the cord to hang it by.

To get started, set the sheep up on its rump (Fig. 4-24). Beginning at the brisket, make downward strokes between the front legs. Methods of taking off the belly wool vary, but usually the first stroke is down the right side. The wool is then broken out and the belly sheared diagonally. Use the left hand to stretch the skin by applying gentle pressure in a direction to prevent wrinkles where

Fig. 4-26. Set the sheep on its rump with its legs to your right and your right leg between the sheep's.

the blade is going. Be particularly careful around the udder. Don't hurry. Do clean work. Speed will come with practice.

With the belly wool off, step back slightly, and shear out the *crutch*, around the udder and inner surface of the hind legs. Next (Fig. 4-25) shear the outer side of the left hind leg. Putting the left hand on the stifle joint will keep the leg in position. In this position, shear over the rump and try to get as far as the right hip bone. The top of the head may now be sheared.

Set the sheep on its rump (Fig. 4-26) with its legs to your right

Fig. 4-27. Lay the sheep down on its right side.

Fig. 4-28. Shear the right side on a diagonal.

and your right leg between the sheep's. With the sheep's head turned back to tighten the skin over the throat, shear from brisket to lower jaw. Break this wool out. The head can be sheared now or before the neck stroke. Leave the right side of the head until you're ready to do the right side of the sheep. While in this position, shear the left side of the neck and the left shoulder.

Lay the sheep down on its right side (Fig. 4-27). Your left leg must be forward of the sheep's front legs, with your left foot on its side and under the sheep's right shoulder. This is very important. Your right knee rests lightly on the sheep's paunch, gently stretching the skin back toward you. Put your left hand on the sheep's forehead. Shear from rear to front and make a stroke or so past the backbone.

Standing straddle of the sheep, pull the head up and shear the right cheek. Pulling the sheep into the position shown in Fig. 4-28, shear the right side on a diagonal. When you get to the right hind leg, shear it right on out. This leads to the position shown in Fig. 4-29. Beginners may find it best to bring the sheep's head up to maintain better control. Shear whatever is left on the right hip and you're done.

The object is to get the fleece off in one piece, with as few second cuttings as possible. Some random remarks at this point may be helpful:

- Don't hurry.
- In all positions, try to stretch the skin to keep it out of the blades.
- Keep the machine blades down on the skin.
- Try to get a full width of the blades with each stroke.
- If possible, learn on mutton sheep. Suffolks are about the easiest.
- You will inevitably make some skin cuts. Iodine them, and if they don't bleed excessively, don't worry about them. Just make as few as possible, being extra careful around udders and the neck.
- Don't shear rams until you know what you're doing.
- Don't shear rams until you know what you're doing.
- If possible, watch an experienced shearer. Better yet, go to one of the shearing shcools that some states sponsor. It's an excellent way to get started.
- Get a copy of the shearing chart sold by the Sunbeam Company. Also, if you can find a copy, read and study Godfrey Bowen's excellent book. *Wool Away.*

Most breeders are very concerned about skin preparation. Many are interested in tanning their own skins. Either way, double salting solves part of the problem and produces professionally handled skins.

Fig. 4-29. Shear the right hind leg as shown.

If you flush your ewes with a pound of grain per head per day for about 17 days at the beginning of breeding season, you may be able to get more of the ewes to have twins. If you have older ewes that consistently produce a single lamb, you might want to get some new ewes that were twins themselves.

Chapter 5

Goats

Dairy goats can either be the most enjoyable animals you ever raised on the homestead or else the greatest headache. Goats are cute, clever, and charming. They are also agile, destructive, and can wreck the farm if they have the chance.

PYGMY GOATS

There is a new arrival that is causing upraised eyebrows among the old-time goat breeders for he is a goat of unconventional size: a "mini-goat" standing only 18 to 22 inches at the shoulders (Fig. 5-1). This new goat, an import, has been quietly gaining in numbers and popularity ever since 1959 when 12 African Pygmy goats were brought into the United States from Sweden by the Catskill Game Farm in New York and a zoo in California. The small goats have developed a following. This group of people has organized a national association to standardize the little goats in order to keep their size and conformation intact.

Originally the Pygmy goat was used by zoos and scientific laboratories for research or as an exotic pet for status seekers. Because of the small size, though, the Pygmy is now in competition with the larger conventional sized goats for the place of honor on the small farm or homestead. Breeders of the standard size milk goats are taking exception.

Apparently, the main worry breeders of the larger dairy goat have is that the Pygmy will pollute their breeds through inbreeding, resulting in an inferior milk goat. Certainly no one can predict what

Fig. 5-1. Pygmy goats stand only 18 to 22 inches high at the shoulders.

some people will do while experimenting with breeding animals, but serious Pygmy goat breeders are interested in small size and are not about to breed the small goat up by using a standard size mate. Equally serious standard dairy goat breeders are interested in larger goats for more output of milk, so there would be no advantage in mixing the two types of goats.

While it is true that one Pygmy doe cannot produce as much as one of its larger cousins, the quality of the milk is outstanding. The Pygmy has a butterfat content of six to nine percent. The Pygmy also has the advantage of being able to eat any type of food without it adversely affecting the taste of the milk, while most milk producing animals are fed very carefully to maintain the flavor of their product.

The Pygmy goat has many advantages over the standard size goat if the farmer is interested in space, because it can be maintained in a much smaller area than any other goat. Many have called a doghouse with a kennel run home. They eat less and browse more effectively than other goats. They actually seem to prefer brush to lush green (they do, however, get just as tangled on a stake-out rope as the larger goats).

Because of their small size, they are easily handled by children. Controlling a 30 to 45-pound imp certainly has advantages over the 100 to 150-pound standard size playful goat.

Their small size also allows more advantages as a pet. They can be taken in the car and often jump into the car when the door is opened, anxious to go for a spin around the farm or to the feed store. They can be taken into the house to snuggle at your feet in front of the fireplace, be led around on a leash, or go and do most anything your dog can. They are death to blackberry bushes and leaves that drop from your maple tree in the fall. They can be taught to come when called, be housebroken, and perform tricks for tidbits.

ANGORA GOATS

Angora goats are found in many states of the United States, but the heaviest concentration is found in Texas. In an area about the size of South Carolina covering roughly 32,000 square miles in central west Texas, Angoras are found in abundance (Fig. 5-2).

This area has gently rolling hills covered with live oak mottes, some shin oak and other browse, and generally good grass cover. Rainfall ranges from 12 to 25 inches, and the elevation is from 500 feet above sea level closest to the Gulf Coast to 3,000 feet in areas west of the Pecos River.

Angoras do well in this rainfall range. They are best adapted to dry areas that provide forbs and browse, preferring these conditions to grasslands in higher rainfall belts. Because Angoras are adaptable, however, they can be raised in a wide range of conditions, but management must be modified to allow for deviations from optimum rangelands.

Getting Angoras into the United States in the mid-1800s was quite a feat. Some of the early accounts tell of smuggling goats out of Turkey. The result was many small flocks of purebreds established around the United States.

Early commercial flocks of mohair-producing goats were established by repeated crossings of Angoras on grade goats. Each succeeding cross dramatically improved hair-producing qualities until the fifth cross. Following this cross, little change was noted from the purebred Angoras.

Fig. 5-2. The heaviest concentration of Angora goats in the United States is in central west Texas.

Between 400 and 600 pure Angoras were reported to have been imported into the United States. About half came from Turkey and the balance from South Africa.

By 1900, estimates showed that Angoras of all grades totaled about 500,000. In 1965, at the peak of American production, numbers were in excess of 4 million. Currently, we have just over 1 million Angoras in this country.

Shearing weights have peaked at 7.4 pounds per head average. Apparently with the severe reduction in numbers in recent years, producers kept their heaviest shearing goats, and these are making up the foundations of current flocks. The best source of Angoras from a number and quality standpoint is central to west Texas.

Numerous livestock auctions throughout this area offer sheep and goats on certain days during the week. The most active of these are auction sales in Junction and San Angelo, Texas. These sales are sources of commercial goats only.

Purebred breeders are members of the American Angora Goat Breeders' Association whose permanent address is P.O. Box 195, Rocksprings, TX 78880. This breed association will provide a breeders directory on request.

Angora does are usually mature enough to breed when they reach about 60 pounds body weight. It is possible for them to achieve this size the first fall after they are born, but under most conditions it is usually their second fall. Therefore, most commonly, Angoras are two years old when they drop their first kids.

Angora goats in the United States are seasonally estrus. They begin cycling in September, normally, with the most active month being October. This is the month when most Angoras are bred. With a gestation period from 147 to 155 days, most kids are dropped from late February into April. Singles are usually dropped with no more than 10 percent of the does having twins under range conditions. Under good feed conditions, kid crops of up to 150 percent can be obtained. Angora goats respond exceptionally well to good nutrition and produce both more kids and fleece.

Under typical range conditions, one buck will service 30 to 40 females. It is poor practice to depend solely on one buck, due to the possibility of his infertility—either permanent or temporary.

One of the greatest mistakes made with Angoras is the assumption that their nutrient requirements are low (Table 5-1). Angoras require a high plane of nutrition, particularly protein. During late gestation and lactation, Angora does need almost ½ pound of crude protein daily to meet their requirements for production of fetus, milk flow, and mohair growth.

Under range conditions it is difficult for Angoras to obtain this level due to the low protein content of dry range feed available during the season near kidding time. Unless supplemental feed is furnished, Angora does frequently abort and/or abandon their kids. Often the kids are poorly developed and fall easy victim to severe weather or predators. Therefore, it is critical that ample nutrition be available during two months before and at least one month after kidding.

Angoras require clean feed and surroundings. They respond to good care and should not be considered scavengers. They thrive on browse and palatable forbs. Being agile animals, they will feed high up in brush, feeding on the very tenderest of fresh growth.

For supplemental feeding, alfalfa hay, cottonseed meal, and corn are basic feeds on which Angoras perform well. For temporary pasture, small grain fields such as oats or wheat are excellent for Angoras.

Table 5-1. Recommended Nutrient Allowances for Angora Goats.

Class & Weight of Goats	Dry Matter	Total Dig. Nutrients	Crude Protein
Wethers and Dry Goats			
50 lbs.	2.1-2.7 lbs.	1.3-1.6 lbs.	.24-.28 lbs.
60	2.4-2.8	1.4-1.6	.26-.29
80	3.1-3.3	1.6-1.8	.30-.32
100	3.8	1.9	.35
120	4.0	2.0	.38
Pregnant Does (last 8 weeks)			
50	3.1-3.5	1.9-2.2	.34-.39
60	3.3-3.8	2.0-2.3	.37-.40
80	4.0-4.3	2.2-2.4	.41-.43
100	4.6	2.5	.45
Lactating Does (16 weeks)			
50	3.1-3.7	2.0-2.4	.38-.44
60	3.5-3.9	2.2-2.4	.41-.45
80	4.0-4.2	2.4-2.5	.45-.47
100	4.4	2.6	.49
Growing Kids and Yearlings			
20	1.8	1.2	.24
40	2.5	1.6	.29
60	3.2	1.9	.34
80	3.3	1.9	.34
Developing Billies			
80	3.7	2.4	.42
100	4.2	2.5	.43
120	4.4	2.5	.43

The primary product of Angora goat husbandry is mohair. Mohair is the woollike fleece of the Angora. It grows at about ¾ to 1 inch per month, and normal production is about 7 to 9 pounds per head per year.

Mohair differs from wool in its luster and silky appearance. It is stronger, withstanding more abrasion than wool. No fiber will dye like mohair to accept the infinite shades that reflect the sparkle and luster found in the fiber. Its main uses today are for ladies' fashions, coats, sweaters, menswear (including formal dress), upholstery, and casement fabrics.

Mohair is a specialty fiber, one of two which comes from goats. The other is cashmere. Mohair is the only important specialty produced in the United States.

Angora goats are said to be the most efficient fiber-producing animal in the world. Normally, Agoras are shorn twice a year, about every six months. In Texas they are shorn in February and early March which is three to six weeks before kidding. They are shorn again in August, at which time the kids are weaned. In northern New Mexico where the weather is more severe, goats are shorn only once each year—usually April and early May.

One of the greatest causes for death losses in Angoras is "freeze loss." These goats are extremely sensitive to chilling winds and wet weather. Commonly, large flocks of freshly shorn goats are entirely lost when they are exposed to rain and brisk winds. Allowing fresh shorn goats access to sheds for protection for four to six weeks after shearing is the only safe way to prevent losses. Adequate nutrition under such conditions must be provided.

The practice of of leaving a "cape" or unshorn strip of hair 6 to 8 inches wide down the back will offer some protection, but it will not prevent losses. Having the animals in good condition will also help considerably in preventing chilling.

Virtually all of the mohair in the United States is shipped from warehouses in Texas. About 95 percent of the clip is exported through the port of Houston. Large grower warehouses located in various towns south of San Angelo handle mohair for producers. Most of these firms handle mohair on consignment and are very reputable. It is advisable for small producers outside of Texas to pool their mohair for shipment to one of these warehouses to save freight. In other areas, small flock owners may find a sufficient market locally with spinners and weavers.

Producers separate various types of fleeces when the goats are shorn. For example each age group such as kids (first two shearings—one at six months and one at 12 months of age), young goats or yearlings (fleeces produced by goats at 18 and 24 months of age), and adult mohair are sacked and sold separately as each has a different value.

Older goat's fleeces are referred to as spring adult when shorn early in the year and fall adult when shorn in the summer and early fall. Fleeces that are over six months growth are usually referred to as 12 months mohair.

Attempts are made to keep the hair clean and free of grass, burrs, and excess urine stain. If these occur, the mohair may be discounted accordingly. Colored hair and kempy fleeces from common or crossbred goats should be packaged separately. Fleeces are usually sacked in burlap bags.

The most common external parasite found on goats are lice. These can be controlled with spraying. Angoras have to be treated for internal parasites, particularly in areas of heavier rainfall and where goats are forced to feed close to the ground. Tramisol, TBZ, or Loxon drenches will do a good job controlling most common parasites.

Additional information about Angoras is available from the Mohair Council of America, 516 Central National Bank Building, P.O. Box 5337, San Angelo, TX 76902, (915) 655-3161.

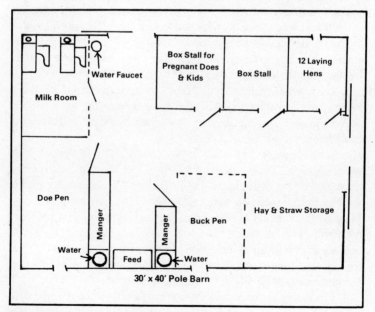

Fig. 5-3. Layout for a goat barn.

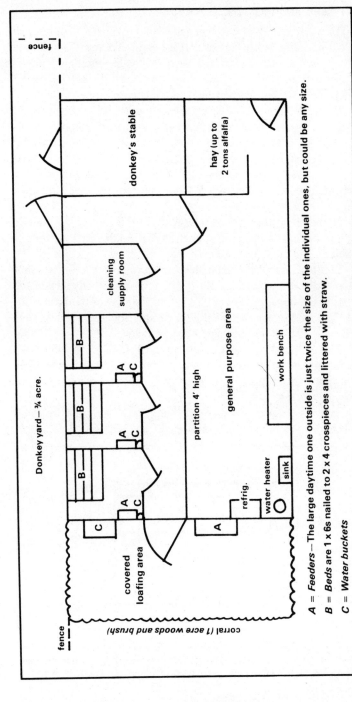

Fig. 5-4. Note the position for the hay feeders.

GOAT BARN AND HAY FEEDER PLANS

This goat pen is 8 by 20 feet and is built of old farm lumber that has been coated with creosote (Fig. 5-3). There is a plywood platform at one end. This was a necessity as the ground sloped at that corner, and the bulldozer didn't get the fill right up to the corner. You can bed with straw and allow it to build to a pack.

On the long side of the barn are keyholes all the way down to the gate. The first three keyholes overlook a platform upon which a washtub for drinking water has been placed. A washtub is the best size for a small herd of goats. A protein block is placed beside the water tub. The goats may eat a lot of the protein or at times they ignore it. The rest of the keyholes face the mangers which are open on the aisle side.

At the end of the aisle is a large divided feed bin. One side is used for storing rabbit pellets, and the other side has a mixture of horsefeed and corn for the goats.

This barn is the usual 8 feet high at the wall, so a ceiling was put in 6 feet high over the goat stalls to keep the goats' body heat in and yet still take advantage of circulation in the rest of the barn (Fig. 5-4). The feeders take very little time to put together, and there is never any hay spilled or wasted on the floor.

The goat stalls consist of one by three slats, spaced 3 inches apart all around so the goats can see and talk to each other when they are in the barn at night. The doors to the corrals are both Dutch doors.

The feeders in the barn are hung onto the slats by large "hairpins" made out of strips of ½-inch iron (Fig. 5-5). They could

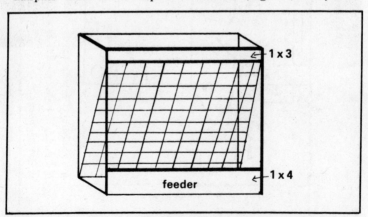

Fig. 5-5. The sides, back, and bottom of the feeder are plywood; the top is open for filling. The rack is two-by-four mesh welded wire left over from the fences.

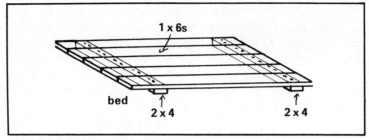

Fig. 5-6. Construction for a bed.

be attached to a solid wall, too. The beds are solid and just high enough to keep the goats up off the cement floor (Fig. 5-6).

EXERCISE YARD

Dairy goats need a fenced outdoor exercise yard for fresh air, sunshine, and exercise that keeps them trim and helps avoid arthritis problems (Fig. 5-7). You save in bedding costs and barn cleaning chores if you have a good yard that encourages the animals to spend as much time as possible outside.

Fig. 5-7. This exercise yard would be the delight of any goat. The cable spools partly buried in the ground are covered with indoor/outdoor carpeting to make them skidproof (photo by Joyce Martin).

A good exercise yard also expands your barn space. You need 25 square feet of loafing pen per mature animal if they are confined to the barn. If the animals can comfortably be outdoors most of the time, 15 square feet is plenty.

Pick a high spot with natural drainage. Fill in low spots. Spread a thick cover of sand on the yard or at least in high-traffic areas close to the barn. Don't use coarse gravel or crushed stone on goat yards—it bruises feet and manure packs down into it. You can't clean the yard without removing most of the stones. It is easy to rake bedding and manure from a sand-base yard if those organic materials start making their own "mud."

You don't need cement in a goat yard. Goats don't weigh much, and sand works fine. The largest commercial goat dairy in the nation, Laurelwood Acres in California, has sand in barns and exercise lots, and they're happy with it.

Goats don't like mud. They won't walk in it if possible. That's a good natural instinct—goats are very susceptible to pneumonia. If the exercise yard is muddy, the goats will stay in the barn and they won't get any exercise. That's bad for the goats.

In hot weather your exercise yard must offer comfortable shade for the animals all day long, so they get out of the barn. They will sleep in the shade.

A good tree is the best possible shade. It is worthwhile to stretch the fencing a bit to include shade trees. Plant a tree or two for future shade over the goat yard.

Protect those trees. Goats will gnaw the bark and kill the trees if they can get to them. Wrap wire mesh no larger than 2 inches by 4 inches around the trunk of every tree to a height of 6 feet (Fig. 5-8). It is better if the tree trunk is on the other side of the fence.

The east side of a building is good shade until late afternoon.

Fig. 5-8. Note the tree trunk covered with chicken wire to discourage nibbling (photo by Joyce Martin).

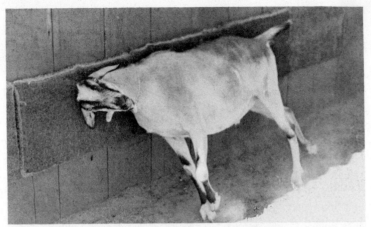

Fig. 5-9. Given a row of six cocoa fiber doormats nailed to the fence, this goat will brush himself along it a dozen times a day and leave the rest of the fence alone (photo by Joyce Martin).

An Indiana goat dairyman built a shade roof at the far end of his exercise yards for his does. The goats liked it and they spent most of their time there, rather than inside the barn, in the heat of midsummer days. A couple of years later, he opened up his fences and gave the does access to a couple more acres of land which had large old oak trees. The does abandoned the shade roof and clustered beneath the trees for summer shade.

Running room is important. If you really want the animals to exercise, they need space to run and bounce and do their typical acrobatics (Figs. 5-9 and 5-10). The pen should be at least 50 feet

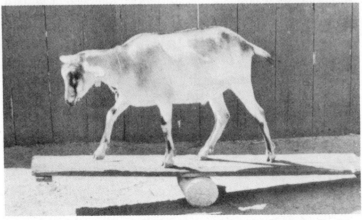

Fig. 5-10. Teeter-totter is a game for one or two goats (photo by Joyce Martin).

long in one direction to give the animals that running room.

If you are cramped for space, the goats can survive with a much smaller outdoor pen. They won't get as much exercise as they might need.

Snow and rain will keep the goats in the barn. A wide roof overhang over the pen is a good investment. It keeps rain off a small outdoor section and the animals will at least stand outside on rainy days. Shovel snow away from the big area around doors in the winter. The animals will spend quite a bit of time outside in winter if a building or a solid piece of fencing protects them from chill winds.

FEEDING PROGRAM

Some people simply do not feed their dairy goats enough. But they think they are taking very good care of the animals.

A friend had three grade milking does. They each produced 3 to 4 quarts of milk a day. He was worried about the animals because they were too thin. He wormed them and treated them with louse powder, since he was sure parasites were to blame.

He bought good feed—a commercial horse ration grain mix, and nice second-cut alfalfa hay. It never occurred to us that he was not feeding enough.

Then we happened to visit the barn at noon one day. It was winter, the goats were confined to the barn, and the mangers were empty. That shouldn't happen. Our idea of "enough" hay is so much there is a bunch of stems and stiff pieces left in the manger at the next feeding time. You sweep those out to add to the bedding.

We really didn't think about it then. It took two more casual visits before we realized that the mangers were always empty and if we happened to offer a handful of hay, the goats fell over each other trying to grab it. This was in the mornings, just a few hours after feeding chores. Our own goats would be so full of feed at that time of day you couldn't interest them with fresh rosebushes.

Finally we did some serious talking about the feeding program. He was giving each doe a heaping measure of grain each feeding. When we weighed it out, it was about 1½ pounds of grain. That was enough grain if the animals had all the roughage they wanted to eat.

He was spreading one single flake of hay out for the three does each feeding and presumed that was enough. It wasn't. To make things worse, the manger was too narrow and rather poorly constructed. A good part of even that hay ended up on the floor and uneaten.

Truthfully, hay was hard to find and expensive. He didn't want to waste any. He really didn't want to feed very much of it until he understood how important good roughage is for a dairy goat.

He started feeding hay generously—enough that there was some in the manger all day long. He found some grass hay that was beautifully cured and less costly than the alfalfa and used this to stuff the mangers full after the alfalfa was eaten. He soon realized how much hay was being wasted and took the time and trouble to build an 18-inch wide manger with keyhole-type head openings.

The goats improved a lot. They actually became a bit plump. One doe produced more than 11 pounds of milk a day. The doe kids that year grew a lot better.

The cost of hay went up also. He ended up feeding three times as much hay as he had been offering the goats before. Remember: you can not feed too much hay.

DRENCHING

Most goats are very particular eaters. While their ability to discern strange tastes and odors usually works to their benefit, it can be frustrating to the goatkeeper who is faced with the chore of worming or medicating a goat. It is often necessary to resort to drenching—a method that can be hazardous. If not done properly, material can reach the animal's lungs, causing gangrene and/or pneumonia, resulting in death.

When the animal swallows, the entrance to the windpipe, the *epiglottis*, closes voluntarily. If, however, the animal doesn't swallow voluntarily when medicine is poured into its throat, the medicine will go down the windpipe instead of the gullet, entering the lungs rather than the stomach. Oil drenching solutions are especially hazardous in this situation as the oil will not evaporate like water if it should get into the wrong place. Also due to the texture of an oil drenching solution, the animal can be fooled while swallowing, and it can slip into the lungs. A good idea is to mix oils like mineral, linseed, and castor with buttermilk and then drench.

Drenching can be performed safely if proper precautions are taken. Some worming boluses can be powdered in the blender and mixed with a liquid. This makes an easier and more effective way of drenching than does using a balling gun. Several emergenices can be handled by drenching. Bloat requires an immediate drenching so be ready. When administering antibiotics, it is recommended to drench with 4-ounce doses of buttermilk twice a day to keep the rumen active. Scours and poisoning are both treated by drenching.

There are a number of things you can use to drench. Probably the most readily available one would be a soft drink bottle. You can also use a poultry baster or a rectal or ear syringe, but the latter is often so soft that a goat can clench it shut with its teeth. A 4-ounce dose syringe is also suitable, but they are somewhat difficult to find.

Be certain to mix drenching solutions accurately, measuring the medicines very carefully. Shake the container occasionally to keep the ingredients mixed. Never guess at dosage quantities.

It is very important that the animal be held properly. A second person or a goat squeeze is helpful. Stand astride the animal's shoulders and reach around with the left hand to hold the muzzle and beard. The beard is an extremely handy protuberance on many occasions. For unknown reasons though, some people trim it off.

Restrain the animal and elevate the head slightly above level position, making sure to keep the nose below eye level. The higher the elevation, the greater the danger of strangling or having the mixture enter the windpipe. The drenching instrument can be put into the side of the mouth at about a 45-degree angle with the tip of the tube resting on the middle of the tongue. Some people prefer to put the tube in frontally, with good results. The medicine should flow slowly and naturally in this position. Never give too much liquid at one time. An ounce is about the maximum amount a goat can swallow at one gulp without strangulation or waste of the solution. Pay close attention that the animal swallows the solution and breathes again before the next amount is issued into the mouth. The medicine can be poured or squirted slowly and gently in intermittent small amounts without ever removing the dosing tube. If the goat coughs, release at once until it can clear its mouth and regain composure.

In case of coughing or strangulation, free the goat's head. Never pull the tongue out of the mouth as it needs to be free for the natural reflex of swallowing. If more than a pint of liquid is given as a drench, allow 10-minute rests between doses. Keep the drenching equipment clean. In emergencies such as bloat there will not be enough time to get ready. Beware of drenching an animal with milk fever or any ailment that might cause paralysis of the throat, plugging of the esophagus, or muscular choke. Also, do not drench animals in a comatose state. Drenching is not only dangerous in such cases, but is usually fatal. Do not drench an animal that is lying down except in the direst emergencies. Keep the nose below the eyes. Drench slowly—the goat's swallowing should be your guide. When drenching, go slowly and remain absolutely calm.

WINTER MILK SUPPLY AND LIGHTING

Everyone who has ever kept dairy goats for a family milk supply knows that a winter milk shortage is a normal event. Milk production drops, often rather quickly, in the fall. The highest milk production comes the first few months after a doe gives birth, and goats are seasonal breeders that prefer to produce their kids in the spring. Kidding normally starts in March.

Goat owners have used a variety of management methods to get winter milk. Heavy producing does can be milked through for long lactations of two years and even several years at times.

The standard management routine is a 10-month lactation followed by a two-month dry rest period, then the doe kids again. The doe is rebred seven months after she freshens. The usual breeding season is September through December; other months are possible but unreliable, so the doe is bred year after year in the fall and then goes dry in winter and comes fresh in early spring.

Goat raisers have found that the very good producing does, those animals which are still giving 4 to 5 quarts of milk a day seven months after freshening, don't need to be bred again that year. They can be milked through until the next breeding season. In some cases a doe has been milked continually and without breeding for three, four, and even as much as six years.

This doe won't grace your milk pail with 6 quarts a day in that long lactation—3 to 4 quarts daily is quite good. Her yearly production total will be less than if she were bred and freshened regularly, but the steady milk supply through the winter is valuable on the homestead. Too often high peak production after freshening is almost surplus milk that's fed to the hogs.

Many dairy goat does do not have the genetic capacity for sustained high production over a long time. Climate, feeding, and management affect this, too. The long lactation reports tend to come from southern states. In the north, severe winters and available winter feed limit milk production.

There are breed differences, too. The very popular Nubians with their rich, sweet-tasting milk are usually (but not always) not dependable for long lactations. The Swiss breeds (Saanen, Alpine, Toggenburg) do better in most cases.

Goat owners usually try to get some does bred "out of season" so they give birth and come fresh in the fall and early winter. Usually this means breeding pens. The dry yearlings are penned with a buck starting in May. Some goatkeepers insist the results are better if there are two bucks in the pen. They compete a bit and their aggressive, garbling breeding behavior helps bring does in heat. Yearling bucks seem to perform better than older bucks.

Older does would give the same results in breeding pens, but the older does are usually milking. It is a logistics headache to move them from pen to milking parlor twice a day, and buck musk on the doe might (this is questionable) affect the flavor quality of the milk. It is easier to keep dry does in those breeding pens.

Hormone injections have been tried to bring does into heat out of season. The injections are not reliable. Results are erratic.

We've got something better than hormones now. We have definite evidence that you can use lights to get does bred out of season quite reliably.

If your barn is wired for electricity, you can treat the animals to long "light days" in the fall by using a not costly chickenhouse timer to turn the lights on and off on schedule. And you probably don't need a zillion dollars worth of fancy intense fluorescent light fixtures to accomplish the deed.

Lights can also take care of the usual sudden milk production drop in the fall (Fig. 5-11). Goat behavior is controlled by the hours-per-day of light they receive. Light is what triggers the breeding/rutting season in the fall. It also affects milk production in the fall.

Goat dairymen have tried the 16-hour light days in November and December, and they're convinced it has helped keep up production. In the northern states it is common for does to dry off rather quickly in early fall when the days get short. Goatkeepers usually blame that on the weather or the fact the animals are off pasture and

Fig. 5-11. Lighting helps keep milk production of goats up in the fall.

eating hay. The real reason is probably the shortened hours of daylight.

The most impressive thing that those barn lights can do is change the normal breeding behavior of goats. They can guarantee you will have does in heat in April and fresh does with a heavy milk supply in September to carry through the winter.

The system for goats was "invented" in Wisconsin by commercial goat dairymen who needed winter milk for their market. Harvey Considine studied the USDA research work on photoperiodic stimulus to change breeding behavior on sheep and goats some 20 years ago. At that time, the animals were being confined to dark barns to shut off summer daylight. It worked in the experiments, but the idea was totally impractical on a working goat farm.

It is difficult to keep a building totally dark. The animals would get too hot shut in a barn in warm weather.

Considine decided the same results might be possible by adding light in the winter instead of shortening the light in the spring. He used timers on his barn lights to create 20-hour light days in January and February, then shut off the timers in March to drop the light stimulus to about 14 hours per day. This duplicated the light change from summer into fall, when the normal rutting or breeding season occurs.

The system worked. Considine was able to get does bred in May and June and had fall fresheners to keep up the dairy's winter milk supply. The percentage of successful out-of-season breedings varied from year to year. Often less than half the does that were in spring breeding pens conceived and freshened in the fall.

Another commercial dairyman, Paul Ashbrook, studied Considine's system and decided the principle was good, but more light was needed. There weren't enough lights in the barn to be a really effective stimulus. In his barn, Ashbrook installed fluorescent lights and used enough of them to make the place almost as bright as a supermarket.

He decided he needed 1 foot of fluorescent bulb for each 10 to 11 square feet of floor space in his barn with a 9-foot ceiling to get sufficient light to the goats' eye level. The extra lighting did the trick. Each year from 75 percent to 95 percent of the does that Ashbrook puts in the breeding pens in May and June have freshened in the fall. He points out that you don't get any better results from breeding in the normal fall season.

Ashbrook exposes his animals to 20 hours of light a day for two months, from January 1 until about March 1. Then he shuts off the timers and the goats get 12 to 14 hours of light a day, depending on when chores are done in the building.

About six weeks later, in mid-April, the does will start coming in heat. They appear to cycle through two or three heat periods, Ashbrook says. The heat periods are extremely short, he found. The doe may be in heat no more than three or four hours. It is important to have the does penned with a buck for breeding. The buck must be exposed to the same light stimulus so he will be ready for breeding.

This commercial dairyman is using lights to breed one-half of his doe herd to freshen in the fall. He has found that when you use enough lights, the system is completely reliable.

The idea of photoperiod light stimulation works for commercial goat dairymen, but is it a practical idea for a homesteader who is producing a family milk supply? Samuel and Maureen Huber of Crockett, Kentucky, say "yes." Their Saanen goats have never been willing to breed out of season. This year, thanks to lighting in their small 20 by 30-foot barn last January and February, they had four does freshen in September and a winter milk supply for the family of eight was assured.

The Hubers didn't have fancy wall-to-wall fluorescent lights. They had one single 100-watt bulb hanging in the middle of the doe pen. It was attached to a chickenhouse timer.

The Hubers discovered the lighting by accident. Their laying hens are penned in the same building, and two years ago they installed a timer and an electric light in the chicken pen to keep up winter egg production. The bulb happened to burn out in March and wasn't replaced. They were surprised when a doe came in heat and was bred in April—none of their Saanens ever bred that late in the season.

They then read the discussion about photoperiodic estrus induction in goats and guessed that the light might account for their fall freshener. Last winter they moved the light bulb to the doe pen and set the timer for 20-hour light days in January and February. The buck is penned in the same building, so he got the light treatment also.

The Hubers penned four dry does with the buck in April and hoped some of them would be bred. All four conceived and freshened.

Their modest lighting system was certainly easier to install than the 30-foot-long stretch of double fluorescent bulbs Ashbrook's lighting formula would call for in that barn.

Ashbrook pointed out that no one really knows how much light is necessary. The lighting he used has given him reliable and dependable results year after year.

"If one 100-watt bulb worked for that herd, that's great," Ashbrook said. "But I'm not sure it would be enough light for good results in every case. It might be adequate one year, but the next year all the does wouldn't come in heat. It may be sufficient for those Saanens but not enough for a different group of animals.

"Until we have been able to test small amounts of lighting, I would advise people that more lighting will usually be more dependable."

BREEDING BASICS

Here are some basic tips for beginners who will be trying to get their goats bred this fall. Generally, the breeding season starts about September and lasts until January. Breeding becomes more of a hit and miss proposition by January. If you need milk for the homestead, don't wait too long to get that doe bred.

Get your does in breeding condition: not too fat and not too thin. Milking does often look skinny to the novice. Their ribs stick out, and there's a hollow in the belly in front of the flanks. To the experienced eye those wide-sprung, wide-spaced ribs are part of something called dairy character. That big hollow is expansion space for the barrel below, which is designed to hold lots of roughage.

If your doe is more than a year old and not milking, she may be too fat. Fat does are hard to get bred, and old fat does are very hard to get bred. You may have to put her on a crash diet—not more than ½ pound of grain daily plus nice leafy hay.

If you are feeding grain and good legume hay or pasture, you probably don't have overly thin animals unless you have parasites in your herd. Check for lice and worms. (A vet should do a fecal test if you suspect worms.)

If you don't have your own buck, by this time you should have made arrangements with a buck owner. Find out what the breeding fee is, if they accept does for boarding, and what the boarding fee is. Have their telephone number handy so you can tell them you're coming when the time arrives.

How do you tell when a doe is in heat? Watch for unusual behavior—bleating, nervousness, indifference to food, or a mild case of stupidity. She may fight with other does. She may try to ride them, or they her. A few milkers drop drastically in production when they're in heat. When you stroke over the hips and backbone, the tail will probably wag furiously. The vulva will probably be slightly swollen and pinkish in color.

If you think a doe is in heat, make a note of it in your record book. She may stay in heat several days: record every one. Chances are in the beginning you will be wrong, but the records will teach you to spot genuine heat cycles pretty quickly. If the doe was really in heat she will be back 21 days later. (Some does have 18, 19 or 20-day cycles.)

In some years does will start coming in as early as August, while in others they might not show obvious signs of heat until October. In fall a doe might stay in heat for four days, while in winter the same goat will only be in heat two days. Records will help pinpoint these variables.

Breeding on the second day of estrus provides the best chances of having a doe settle. If she's bred, that's it. Watch the next heat cycle just in case. If she comes in again, back to the buck.

KIDDING

If you're a homesteader with one or two goats, then, like us, you look forward to a brief but exciting kidding season. Unless you have a whole herd of goats all kidding within weeks of each other, the tendency is to make each birthing a big production. Attempts to help quite often just interfere in what is a pretty natural event.

Given a healthy doe and a normal delivery, there should be no reason for you to scrub up and act as midwife. On the other hand, your goat is a domesticated animal dependent on you for her welfare. If at all possible, you should be there during the delivery to see that it is a normal delivery, to offer assistance if needed, to see that the kids are breathing and cleaned off, and to watch one more miracle of life.

The gestation period for goats is 150 days. A doe may deliver anywhere from 143 to 157 days after conception. Knowing when your does are due doesn't actually give you a scheduled appointment. For example, two of our goats were bred 12 hours apart and delivered eight days apart. Nevertheless, we mark 150 days after service on our calendar and about two weeks before that date we prepare for the kidding.

First we gather together the following: a clean bucket for hot water, an antiseptic soap for washing hands, scissors and a couple 12-inch lengths of string to cut and tie the umbilical cord is necessary, a couple tablespoons of iodine in a baby food jar for disinfecting the kids' navels, clean old towels or rags, two pop or beer bottles with lamb nipples, a funnel, a milk pail and udder-washing bucket, and a box large enough to hold two kids, half-filled with clean straw, to keep them warm and safe until ready to be fed. Everything that lends itself to it is sterilized with boiling water. All else is as clean as possible.

Some people prepare a special kidding pen for the doe. Our does stay in their individual stalls. We lay down fresh bedding with an extra pile of it in the corner to be laid down when the doe goes into labor.

Since we separate the kids from the doe right away, we also have the kids' stall to clean and get ready. With those three chores done, we're ready any time she is.

About two weeks before kidding, the end of the doe's spine will seem to rise. The muscles below the last few inches of spine before the tail are becoming relaxed in readiness for kidding. The closer to delivery, the greater grows that hollow beneath the spine (Fig. 5-12).

Her udder will begin to fill up now, too. Sometimes the udder will look uncomfortably full, but feeling it usually assures me that there's room to spare. Some people suggest that a turgid, shiny udder should have a bit of the colostrum milked out. If you do so,

Fig. 5-12. This goat's labor is near. She's restless, the end of her spine is up, and her udder is full (photo by Bill Allred).

Fig. 5-13. With a soft grunt she bears down as a contraction reaches its height. Delivery should occur within an hour (photo by Bill Allred).

freeze the colostrum. You may want to use it later.

The signs that tell you today's the day are: the doe's refusing her grain; a white vaginal discharge, quite thick and very obvious; and lastly a great restlessness—lots of standing up and laying down, baaing, looking at her back end as if for her kids, and pawing the ground. (Loss of appetite can occur for a variety of reasons and in and of itself need not indicate the beginning of labor.)

When the doe is visibly straining and grunting, she is going into the final stage of labor (Fig. 5-13). She should produce a kid in half an hour, though an hour or so is not particularly alarming, especially if you can see that her efforts are productive.

First, the external genital area will begin to bulge and stretch with each contraction. Next the opening with each contraction. Next the opening of the birth canal widens, and a shiny membrane (the amniotic sac) will appear and recede with the contractions (Fig. 5-14). Our does have always delivered with the sac intact, but it's not uncommon for this "bag of waters" to break now or even earlier in the labor.

Fig. 5-14. The amniotic sac appears and recedes within the contractions (photo by Bill Allred).

In a normal presentation, the front hooves appear first with the head between or resting on top of them (Fig. 5-15). A few more good pushes delivers the head (Fig. 5-16). At this point the doe will rest. She will cease bearing down for a few moments, and she may get up (if she isn't already standing) and walk around for a bit. The amniotic sac should have broken by now (Fig. 5-17). If it hasn't, you might want to break it. The liquid will be lubrication for the final push.

The important point is that once the bag is broken and the kid's face hits the air, it's going to try to breathe. Mucus can clog the mouth and nostrils and should be removed. Usually the kid will sneeze and cough and take care of it itself (Fig. 5-18). If not, tickle the nose to get the kid sneezing or wipe its nose gently with a soft cloth.

When the doe's contractions resume, stand by—one or two good pushes will send the rest of the kid into a dive to the floor (Fig. 5-19). If you can get there fast enough, try to break the kid's fall by letting it pass through your hands on the way down. It's no great

Fig. 5-15. The kid's hooves and muzzle appear (photo by Bill Allred).

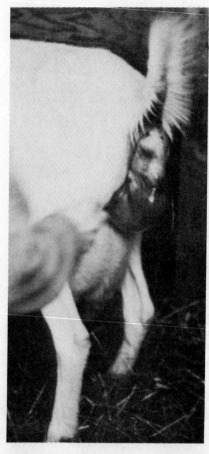

Fig. 5-16. The head is delivered with the water sac intact (photo by Bill Allred).

tragedy if it lands directly on the soft bedding.

The first thing to do is again wipe off the kid's nose and wipe out the mouth with a clean cloth or your clean fingers. The second thing to check is the condition of the umbilical cord. Usually it breaks as the kid is delivered. If it hasn't, it will be necessary for you to cut it. Make sure the blood is drained from the cord. Blood leaves the cord naturally within minutes of delivery. Then cut it about 8 inches from the kid's body (where it would have broken naturally). Some people will tie a string close to the stomach and cut off the excess cord and string ends an inch from the knot. In any case, it is important to disinfect the navel with iodine. First, put the kid down in front of its dam for a little cleaning up.

Using her tongue, a doe can wipe off her baby faster than you can with six large cloths (Fig. 5-20). After she's scrubbed her down

down to the fur, it's time to disinfect the navel. Kneeling down with the kid upright in front of you, cup the navel and the cord with the mouth of the baby food jar with iodine in it. Holding it firmly in place, roll the kid over onto its back across your lap and then upright again. This douses the navel very thoroughly. Some cotton dipped in iodine and gently rubbed on the area also would do the job.

After the disinfecting, the kid can go into her box. The second kid, if there is one, comes fairly soon and pretty quickly, with little straining by the doe. The same cleaning up and disinfecting goes for kid number two.

With the kids temporarily set aside in their box, we can give the doe some attention. She's offered a tall bucket of water that's as hot as a cup of coffee, and she'll usually drink a lot. According to some folks, the water's heat replaces that which she lost when the

Fig. 5-17. The sac breaks (photo by Bill Allred).

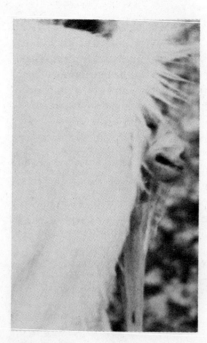

Fig. 5-18. The kid snorts and squeezes (photo by Bill Allred).

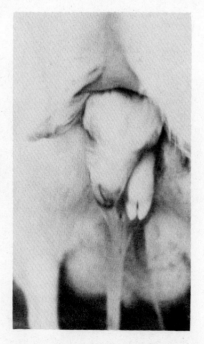

Fig. 5-19. After the head is delivered, the doe pauses in her efforts before delivering the rest of the body (photo by Bill Allred).

Fig. 5-20. The doe cleans the kid quickly and efficiently (photo by Bill Allred).

kid left her body. Other people insist on adding a little molasses or salt or vinegar or even a bit of tea to the water.

Bring the doe to the milking stand to milk out some colostrum. If your kids are to stay with the doe, they can go back in her stall now. Make sure you see them nurse before you leave the barn.

Milking or nursing the doe right away is also to the doe's advantage since the action causes the uterus to contract and expel the afterbirth (Fig. 5-21). Don't try to empty the udder. For their

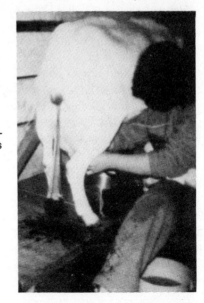

Fig. 5-21. Not all does expel the afterbirth as quickly as this goat does (photo by Bill Allred).

first feeding the kids only need 4 to 6 ounces of colostrum, so you're not going to need a lot. Never thoroughly milk out the udder for the first two days after delivery. The colostrum is rich stuff and trying to replace it might cause the doe to seriously deplete her own supply of vitamins and calcium.

At the milking stand, the doe is offered her usual grain mixture. For this first time on the stand, it's good to spend a little extra time washing her udder with a warm cloth and drying it carefully. The whole ritual helps to bring down the milk and start the uterine contractions.

After she's milked, the doe can go back to her stall where we've added fresh bedding and put fresh hay in her manger. Her job is over, and we have only to be sure that the placenta or afterbirth is delivered. Again with our girls, it's usually passed by the time we're ready to leave the barn. For some does, it takes several hours to deliver. Never pull on the placenta in an attempt to help its delivery. If, after 24 hours, there is no placenta forthcoming, something is wrong and a vet should be called in.

The placenta has a certain nutritional value all its own, and it's perfectly appropriate for the doe to eat it. Some folks take it away saying that the goat will lose her appetite if she eats it. It's quite possible that the afterbirth is all the nourishment she needs for awhile and that's why she's not interested in eating. For a few weeks after delivery, it is normal for the doe to have a bit of a bloody discharge that gradually lightens and lessens as the days pass.

Right after milking the doe, pour the colostrum into the kids' bottles and set them in a still hot udder-washing water to keep warm. Should it cross your mind, don't try to filter the colostrum—it's too thick to go through the filter.

It is said that the benefit from all the vitamins and antibodies in the colostrum is greatest when fed within the first hour after birth. Be that as it may, it's nevertheless important that the kids get as much colostrum as they want at the first feeding. We try to have 6 ounces in the bottle to allow for spilling and dribbling.

They don't always catch on right away, and the first feeding requires a little coaching. Put your hand under the kid's face so that the chin rests in the palm of your hand. Curl your fingers around the mouth and pry it open with whatever finger fits and push the nipple in. Keep your hand there to cradle her jaw, and keep her and her bottle steady and in a straight line. She'll usually suck eagerly.

After they're fed, the kids can go to their own special draft-free stall. Our kid stall is rather roomy, so we arrange an L-shaped area out of two bales of old hay and set them in the corner of it for a cozy night's sleep or at least four hours of rest.

RAISING KID GOATS

There is no right way to raise kid goats. Every breeder has some different methods, and they can all work well.

First, you must decide if you are going to let the kids nurse their dam or raise them by hand—feeding milk from a bottle or pan. Some people let the kids nurse a few days or a couple of weeks and then change to hand feeding.

If you hand feed, there's great variety in how much milk the kids get and for how long. Daily milk feeding ranges from a quart a day per kid to almost a gallon a day per kid, depending on the breeder. Some folks wean their kids at two months, and some keep feeding milk until six months or older.

Let's look at natural nursing. It seems easiest to simply let the kids nurse their dams. That's "nature's way," with no bottles to wash and no milk to warm. It certainly saves a lot of work, and you can raise some fine kids.

There are nursing-and-bottle systems, but there are problems. A fresh doe might produce anywhere from 2 quarts of milk a day to 6 quarts a day or more. She might have anywhere from one to four kids. The amount of milk is not tuned to the number of kids. Sometimes the doe won't have enough milk for all her offspring.

If you really want to raise those kids, you have to watch carefully to be sure they're getting enough to eat and they are all eating properly. The first week is the critical time.

Kids will slip through fences that hold adult goats nicely. If you have kids running with the does, you will probably have trouble keeping them out of the mangers and in their share of the barn. This may test your carpentry abilities.

When kids are nursing a doe, you still milk out the doe once a day for leftover milk. After a while, you'll know if there is leftover. High-producing does can feed a kid or two and also supply some milk for the household. Occasionally does give so much milk they must be milked. When the kids are weaned, the good doe will keep on producing for several months.

Kids that nurse tend to be wild and not friendly. This bothers a lot of goat raisers who have worked with hand-raised kids. Those animals are wonderfully tame and human-oriented, but it varies with the particular goats.

Some kids refuse to be weaned. Occasionally you end up with a yearling milker that is still trying to nurse her dam. We let butcher kids nurse because it is no-labor and they leave at two months of age. If we happen to let a doe kid nurse, she will be moved to a separate building at weaning and won't see her mother again for a year. By then she has forgotten.

Some breeders let the kids stay with the doe to nurse for three days. That gives the kids frequent feedings of colostrum milk, and that milk isn't kept for human consumption anyway. The regular kid attention might be helpful when the doe's udder is congested, but be careful. If you have a real hard udder with no milk, the kids will merely starve to death.

Some kids will nurse for three days and then accept a bottle easily. We know goat breeders who use this system regularly with no problems. They have unique kid goats. The little caprines we have met in 25 years of goat raising usually would rather starve to death than drink from a bottle once they have found the dam's udder. That's why we dropped this system a long time ago. The kids would not cooperate.

Give the newborn kid a first feeding from a bottle. Then teach it about the udder. Let the kid nurse, but once each day feed some milk from a bottle. When you change to bottle-only, there should be less trouble.

Hand feeding is popular. Many breeders feel this is the only way to raise kids.

You can use a pop bottle with a lamb nipple, some variety of "lamb bar" milk feeder, or teach the kid to drink milk from a pan with the very first feeding. Put warm milk in the metal cup for the first lesson and gently push the little head in.

How much milk should you feed, and how often? Three or four feedings a day the first two or three days and then twice a day feeding is fine. Our kids get extra feedings the first day and then they are part of the twice a day A.M.-P.M. chore schedule.

We feed goat milk. In the winter when we need that goat milk for customers, we have had fine results from cow milk out of the neighbor's bulk tank. Take a week for the switch, using a little less goat milk and a little more cow milk each feeding. The kid pen will smell like calves. The different milk changes the urine.

Milk replacers give mixed results. Stick to goat milk the first few days. Then make a very gradual switch. Opt for a lamb milk replacer if you can get it.

How much milk? You can go several ways.

A kid goat will do just fine on a pint of milk fed twice a day. If you provide nice hay and free-choice grain in proper feeders starting at a few days of age, the well-growing kid can be weaned at two months. Proper feeders hold the feed fresh for the animals, and they can't climb into them. Milk can be merely a "baby food" the first few weeks until the little goats are eating hay and grain well (at least ½ pound of grain a day) and have developed a proper rumen.

Technically, grain is cheaper kid feed than milk. The sooner the kids are eating lots of grain and hay, the better.

That presumes you have an economic use for the milk the does produce. You can sell it or use it for your household.

Some breeders feed kid goats lots of milk for many months—2 quarts a day, or even more, four months, or even six months or longer. The system grows good kids. It is the only way to grow butcher kids that need a lot of flesh at a young age. Breeding stock kids can develop more slowly and catch up later nicely on good regular feed.

It is a matter of alternative feeds and the cost of those feeds. If you have plenty of surplus milk, it is cheaper kid feed than buying grain, possibly cheaper than extra hay for the kids. If you do decide on heavy milk feeding for many months, give the kids hay and grain, or hay at least, free choice, so they can develop the internal capacity to handle these feeds.

You can feed milk via bottle/nipple or pan. Long ago we used pans and liked them. There's an old idea that milk from a pan goes to the wrong part of the kid's stomach. We never saw any difference in growth rates and basic health.

Nubian kids' ears got milk-soaked in pans. Kids put their dirty hooves in the pans. Bottles were cleaner if more work. We opted for bottles. Now the "lamb bar" nipple-and-tube milk feeding systems have greatly cut down the basic work if you have a lot of kids to raise.

DEHORNING

If you've got a goat to dehorn, the best thing is to take the goat to an experienced veterinarian who knows how to handle the situation. If you're determined to do the job yourself, you can use rubber bands or a saw. For kids, you can use a dehorning scoop or take the hot disbudding iron to horns that aren't more than 2 inches long.

Many goat raisers have used rubber bands and the method works. Few really like the rubber bands. They cause too much pain for too long. Usually the horn gets broken off and you have a lot of bleeding.

You can use special elastrator bands or just heavy bands from the stationery store, stretched and wrapped as tightly as possible. Get the band as close to the base of the horn as you can.

Some say you should notch the horn with a file so the band stays in place. A couple of persons said the notch must be below the skin line. Other folks didn't bother with notches and the horns came off, so it probably depends on the size and shape of the horn.

Wrap adhesive tape over the band so the goat won't break it by rubbing the horn. The band slowly cuts through the horn. In about six weeks the horn falls off.

Dehorning Saw

A dehorning saw is usually a bloody process and can be very painful for the goat—and probably for the goat raiser who has to do it as well. Many breeders consider this best if you've got horns to take off.

A veterinarian can reduce the pain with local anesthetic or tranquilizers, or can use general anesthetic. Some people think there is too much danger with these pain-killing methods and prefer not to use them at all.

The typical objection to sawing off horns was too much bleeding. A Texas breeder used a surgical saw on two does, but stubs of horns grew back. A New Jersey goat raiser had the horns sawed off a yearling, and the doe got maggots in the wound and was quite ill.

A Rhode Island breeder doesn't want to be the guy who gets stuck holding the goat again. He helped his vet cut the horns off a yearling wether and says the animal would not come near him for more than a year afterwards.

Another eastern breeder says her two goats reacted in similar fashion to dehorning. It was months before they would trust her again.

Dehorning Scoop

Several breeders suggested a small calf dehorning scoop to remove horns from kids. The device has a pair of handles and you squeeze them to make the cutter head dig in and slice out horn and root.

"I've had good success with the scoop on horned kids," reports a New Jersey breeder, "but it's a bloody mess and hurts . . ."

An Iowa breeder uses the scoop on all kids and prefers it to any disbudding method. "We let the kids horns grow to 1 inch high," she reported, "then scoop them out and put on blood stopper."

"Use calf gougers and make sure to cut off the horns below the hairline," suggests a California breeder. "Pull the veins to stop bleeding and put on blood stopper and something for flies. Feed on the ground for a while so they don't get hay in the holes where the horns were."

An Indiana breeder says that the closer you cut with the scoop, the better. If you think you went so deep you'll probably kill the goat, chances are you did it just right.

A Pennsylvania breeder had bad luck with the scoop. They tried it on 12 kids once and lost three of the young animals to meningitis.

Telling If a Kid Will Have Horns

When a kid is born, it's easy to tell if the kid is horned or naturally hornless. If neither sire or dam were hornless, the kid will be horned.

A horned kid will have a curl of hair over each horn button. The skin will be tight over the skull there. If you clip the hair and look real close, there's a small shiny patch of hairless skin.

The hornless kid has straight hair over the poll. The skin moves easily.

If you're in doubt, keep checking and the tips of horn will start coming through—within a week for Swiss breeds, maybe two weeks for Nubians. Bucks have more horn sooner than does. A hornless head does get small bumps on it, and some people have trouble deciding for sure.

If you are really in doubt, disbud the kid anyway. If the kid was hornless, all you will do is eliminate those "hornless bumps" the mature animal has—and sometimes they are big and ugly anyway. The only problem you may cause is genetic. You could end up breeding hornless animals together by accident if one of them was disbudded. This doesn't always cause hermaphrodites.

Dehorning Mature Goats

Some people think you can't cut horns off a kid or the horns will grow back. Many breeders reported good success on kids. Some people have had the horns return even on mature goats.

A New York breeder took the horns off a four-month-old doe kid with rubber bands, and it worked fine. "You'd think the kid was naturally hornless," he reports.

A Missouri breeder says horns will grow back if you use rubber bands on kids under a year of age. She cuts the horns of kids with a hacksaw.

A California breeder tried rubber bands on a seven-month-old wether. The horns continued to grow in an ugly, deformed manner.

A Michigan breeder had dehorned numerous goats with a variety of methods. She says there is often some horn growth afterwards no matter what age the animal or how well the job is done.

An Iowa breeder uses a hacksaw or a dehorning saw and says the horns will not grow any more if you take a ring of hair with them. Like a fingernail, she says if any root is left a scur will grow.

"If you don't cut deep enough and get a scur, use an electric dehorning iron on it and it will quit growing," suggests a California breeder.

An eastern goat raiser has had her vet cut horns off several kids, and the horns never grew back. Another eastern breeder has used a wire saw twice to cut scurs off his buck, but the scurs keep growing back.

"The vet cut the horns off our first goat and they grew back," reports one New Jersey goat raiser. "She was about two years old at the time."

When horns are cut off kids, the idea, one breeder suggests, is to cut through the skin and down to the skull. At least he thought this was the proper method. Then he saw two four-month-old Nubian does that had their horns cut off too high—at the junction with the skin. The horns did not grow any more.

A Toggenburg buck was disbudded as a kid but grew scurs. Twice the scurs were cut off with a wire saw and grew back. The buck was taken to a clinic for dehorning.

He was given a general anesthetic. The skin was cut from the base of the scurs, and the scurs were sawed off at their junction with the skull. An electric cauterizer was used to stop bleeding. There were no visible ridges of horn left. There appeared to be nothing but bare bone around the hole. The scurs have grown back.

Disbudding Kids with a Hot Iron

Disbudding kids with a hot iron is really not a difficult procedure (Figs. 5-22 and 5-23). A novice could do it. You are not likely to burn so much you injure the kid. Typically the beginner doesn't burn enough and has to go back and "touch up" the job later.

When we started raising goats, the people we bought stock

Fig. 5-22. Trim the head with clippers or a pair of blunt scissors. This allows you to see what you are doing. While you are trimming around the buds, plug the iron in and and let heat for about 15 minutes.

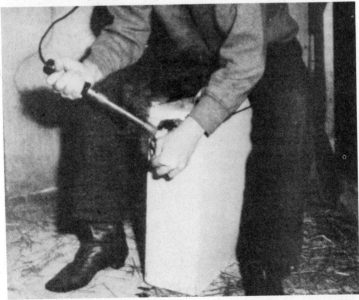

Fig. 5-23. Hold the kid's head firmly in your free hand and burn each button for 10 seconds. Let the head cool down and burn some more if you are working on a good-sized button.

Fig. 5-24. This big Nubian doe is almost too big for the box. Be careful not to pinch a tail when closing the lid.

from used a disbudding iron. At first we took our kids to them. After we watched the process a few times, we bought an iron, built a holding box, and have been doing our own kids ever since with no trouble.

It doesn't seem like much of an operation. The kids squawk and there's bad-smelling smoke in your nostril, but you learn to ignore these things. All that noise tells you the kid is still in good shape. Anyway, they start yelling when you put them in the holding box and you know that isn't hurting them.

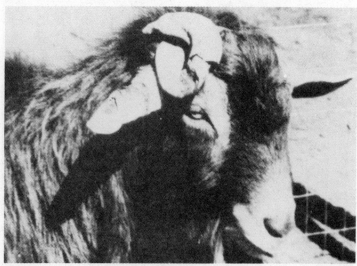

Fig. 5-25. Horn scurs should be touched up with a hot iron as soon as it's obvious that the original burning didn't do the job.

Last winter we met some people who use caustic sticks to disbud their kids. We've never seen chemicals used, and they scare us. These folks have never seen an iron used, and it scared them. It's worth your trouble to go watch a professional disbud some kids before you get started.

A holding box is essential (Fig. 5-24). It lets you do the job with no help at all. It restrains the kid better than any assistant can.

A lot of poor disbudding jobs are due to the fact that the kid squirmed so much the goat raiser gave up before he had burned enough. The box solves this problem.

We like the disbudding iron because you can go back and touch up an inadequate job. Everyone misses some horn root from time to time, especially on bucks. Start checking the kids for pieces of scur at about three months of age (Fig. 5-25). If you find some, heat up the iron and burn the scur off. With a kid this age you can do a lot of burning without harming it.

Some folks use the iron to burn off horns an inch or two long and perform a rather belated disbudding on the horn root beneath. When a goat is dehorned, cauterize with a hot iron. It stops the bleeding and probably gets the last of the horn root.

Hot iron disbudding really is almost painless. The minute you lift up the iron, the hurt is over. The only exception has been the buck kids. We tried to deodorize at the same time. We almost eliminated all the skin on the top of their heads. They were unhappy for a whole day.

Chapter 6

Poultry

The total array of chickens, ducks, geese, and turkeys that we term poultry consists of many shapes, sizes, and colors. In addition to those obvious features, there are variations in temperament, feed efficiency, egg shell color, growth rate, and feather development. With so many different characteristics in the picture, the number of possible combinations is almost endless. Therefore, it is essential to have a standardized set of terms and an orderly system of classification.

Just as we give names to people to identify them, we also use names with poultry, usually by groups of similar individuals. Just as the name of a person calls to our mind that person's particular set of features, stance, and behavior, so does the name of a group of similar chickens call their characteristics to mind. This kind of naming is necessary in order that we understand each other in conversations about them. It is also necessary as we move from one part of the country or world to another and try to relate the merits of the various and sundry fowls.

The first attempts at naming groups of fowls were made in Europe and tended to identify them with the particular locale in which they were developed. Such names as Surrey Fowls, Sussex, and Rhode Island Reds are examples. In some cases, the name of the person who developed them was used, such as the Sebright Bantam that is credited to Sir John Sebright.

As poultry shows developed and people began moving over

greater distances, a more complete and organized system was needed. That was the major reason why the American Poultry Association was organized in 1873. It immediately set about to develop the American Standard of Excellence, later called the Standard of Perfection. This book, which has been periodically revised and expanded, is today recognized worldwide as the accepted guide and system for listing and describing the various kinds of domestic poultry. It is true that some individual nations have their own variations of this, but generally speaking, the "American Standard" is the most universally accepted one.

The "Standard" lists the various chickens as follows:

I. Class (continent or geographic area of origin)
   A. Breed (the general shape, outline, and set of physical features)
      1. Varieties (those different color patterns or comb types that may exist within a breed)

Thus, we have a Barred Plymouth Rock listed as American Class (developed in the United States), Plymouth Rock Breed, and Barred Variety. Another example: Single Comb White Leghorn; Class—Mediterranean (developed in Italy); Breed—Leghorn; Variety—Single Combed White.

When we say a name such as White Plymouth Rock, it immediately calls to the listener's mind a set of physical features, size, shape, color, temperament, egg shell color, etc. It is a convenient and necessary way of keeping the picture reasonably clear because there are more then 350 different combinations of physical features and colors that exist in the poultry world.

Remember that a breed is a group of individuals possessing many characteristics that are passed on to their offspring. Remember, also, that a variety is a subdivision of a breed usually differing from another variety of that breed by only one characteristic. Hence, we have Buff Orpingtons and Black Orpingtons. They are the same size, shape, etc., but one is black and one is buff. They are separate varieties of the same breed. Why do we have so many breeds and varieties? Nature made it possible with the wide genetic variation that exists in poultry. Man loves variation and takes pride in possessing the unusual.

ANCONAS
Class—Mediterranean
Varieties—Single Comb, Rose Comb

Standard Weights:
>Cock—6 pounds
>Hen—4½ pounds
>Cockerel—5 pounds
>Pullet—4 pounds

Use—A small fowl that lays a fair number of often rather small eggs.

Characteristic—Small, active, black with an even distribution of white-tipped feathers (about one in five), Anconas are rather noisy and considered to be nonbroody.

Origin—Anconas take their name from the port city of Ancona, Italy, where they are said to have originated.

General Statement—Anconas resemble Leghorns in shape and size. They are attractive, alert, and good foragers. They were at one time a prime egg producing breed, but today are mainly kept as an ornamental fowl.

BLUE ANDALUSIONS

Class—Mediterranean
Varieties—None
Standard Weights:
>Cock—7 pounds
>Hen—5½ pounds
>Cockerel—6 pounds
>Pullet—4½ pounds

Color of Egg Shells—White

Use—An ornamental fowl with fairly good egg production potential.

Characteristics—Small, active, closely feathered birds that tend to be quite noisy and rarely go broody.

Origin—Developed initially in Spain, the breed has undergone considerable development in England and the United States.

General Statement—Andalusions are a typical example of the unstable blue color we see in the poultry industry. It is the result of a cross of black and white. When two blues are mated together, they produce offspring in the ratio of one black, two blues, and one white. These whites and blacks when mated together will produce mainly blues. Andalusions are beautiful when good, but the percentage of really good ones run rather low in many flocks because of this color segregation. They are not widely bred and never in large numbers.

BRAHMAS

Class—Asiatic
Varieties—Light, Dark, Buff
Standard Weights (Light):
    Cock—12 pounds
    Hen—9½ pounds
    Cockerel—10 pounds
    Pullet—8 pounds
Standard Weights (Dark and Buff):
    Cock—11 pounds
    Hen—8½ pounds
    Cockerel—9 pounds
    Pullet—7 pounds.
Color of Egg Shell—Yellow
Color of Egg Shell—Brown
Use—A very heavy fowl for the production of heavy roasters or capons. Fair egg layers.
Characteristics—Slow moving, docile, very large, presenting a stately appearance. Brahmas do go broody and are fairly good mothers. Their small comb and wattles, together with profuse feathering and well-feathered shanks and toes, enable them to stand cold temperatures very well.
Origin—The ancestry of the Brahma traces back to China, although much of their development took place in the United States between 1850 and 1890.
General Statement—Good Brahmas are beautiful birds. Their large size and gentle nature combined with intricate color patterns make them favorites of many. The Brahma's appearance in the showroom never fails to command the admiration of one and all. These qualities have made them a favorite with showmen and fanciers. The relatively slow rate of growth and long time required to reach maturity have caused Brahmas to be passed by as a commercial fowl in recent times.

COCHINS

Class—Asiatic
Varieties—White, Buff, Black, Partridge
Standard Weights:
    Cock—11 pounds
    Hen—8½ pounds

Cockerel—9 pounds
Pullet—7 pounds

Color of Skin—Yellow
Color of Egg Shells—Brown
Use—Mainly an ornamental fowl, but their ability as mothers is widely recognized. Cochins are frequently used as foster mothers for game birds and other species.
Characteristics—Cochins are extremely docile, stand confinement well, and are relatively quiet. They exhibit extremely persistent broodiness, are good mothers, and are normally intense layers for short periods. Their excessive feathering is rather loosely held, giving them a large bulky appearance. Because of this feathering, it is necessary to clip some of the feathers or resort to artificial insemination to obtain good rates of fertility.
Origin—Cochins came originally from China but underwent considerable development in the United States and are now found and admired in many parts of the world.
General Statement—Cochins are literally big, fluffy, balls of feathers. They are mainly kept as an ornamental fowl and are well-suited to close confinement. The profuse leg and foot feathering makes it desirable to confine Cochins on wet days and where yards may become muddy to prevent the birds from becoming "mired" or collecting balls of mud on their feet.

CORNISH

Class—English
Varieties—Dark, White, White Laced Red, Buff
Standard Weights:
Cock—10½ pounds
Hen—8 pounds
Cockerel—8½ pounds
Pullet—6½ pounds
Color of Skin—yellow
Color of Egg Shell—Brown
Use—Developed as the ultimate meat bird, the Cornish hen has contributed its genes to build the vast broiler industry of the world. Its muscle development and arrangement give excellent carcass shape.
Characteristics—The Cornish has a broad, well-muscled body.

Its legs are of large diameter and widely spaced. The deep set eyes, projecting brows, and strong, slightly curved beak give the Cornish a rather cruel expression. The feathers on the Cornish are short and held closely to the body. In some cases, the feathers may be so short and narrow as to show exposed areas of skin. Because of their shape, good Cornish often experience poor fertility and artificial mating is suggested.

Origin—Cornish were developed in the shire (country) of Cornwall in England where they were known as "Indian Games." They show the obvious influence of Malay and other oriental blood. They were prized for their large proportion of white meat and its fine texture.

General Statement—Cornish are movers; they need space to exercise and develop those muscles. The old males are quite subject to stiffness in their legs if they do not receive sufficient exercise. The females normally go broody, but because of their very minimal feathers they can cover relatively few eggs. They are very protective mothers but tend to be almost too active to be good brood hens. Cornish need adequate protective during very cold weather as their feathers offer less insulation than found on most other chickens. Cornish males are often quite pugnacious, and the chicks tend to be somewhat more cannibalistic than some breeds. Good Cornish are unique and impressive birds to view.

DORKINGS

Class—English
Varieties—White, Silver Gray, Colored
Standard Weights (White):
    Cock—7½ pounds
    Hen—6 pounds
    Cockerel—6½ pounds
    Pullet—5 pounds
Standard Weights (Silver Gray and Colored):
    Cock—9 pounds
    Hen—7 pounds
    Cockerel—8 pounds
    Pullet—6 pounds
Color of Skin—White
Color of Egg Shells—White
Use—A good, general purpose fowl for the production of meat and

eggs. It was developed for its especially fine quality meat.

Characteristics—The Dorking has a rectangular body set on very short legs. It is five-toed and has a relatively large comb, is a good layer, and is one of the few birds with red earlobes that lays a white-shelled egg. Most Dorking hens will go broody, make good mothers, and are quite docile.

Origin—The Dorking is believed to have originated in Italy, having been introduced into Great Britain at an early date by the Romans. Much of its development took place in England where it was acclaimed for its table qualities. The Dorking is one of our oldest breeds of chickens.

General Statement—A good, all-around fowl that is ornamental, unique, and possessed of good economic qualities. Because of their large combs, it is desirable to offer adequate protection to Dorkings in extremely cold weather. Because of their white skin, Dorkings are not as popular in America as they are in Europe.

JERSEY GIANTS

Class—American
Varieties—Black, White
Standard Weights:
    Cock—13 pounds
    Hen—10 pounds
    Cockerel—11 pounds
    Pullet—8 pounds
Color of Skin—Yellow
Color of Egg Shells—Brown

Use—A very heavy meat-type fowl for heavy roaster and capon production. Fairly good layers. The dark-colored pigment from the shanks is prone to move up into the edible portion of the carcass, and this has hurt the Jersey Giant in commercial circles.

Characteristics—A very large, angular bird with single comb and black (with willowish tinge) shanks in the Black variety and dark willow shanks in the White variety. The Jersey Giant normally will go broody, but because of its very large size is not the best choice for incubating and brooding. The Jersey Giant is the largest breed in the American Class. It has a single comb.

Origin—Developed in New Jersey in the late 1800s, at which time there was a demand for heavy fowl for capon production, particularly for the New York market. Size was a prime consideration.

General Statement—Jersey Giants should be extremely large, rugged birds. They are quite angular in appearance and gained considerable popularity for a time but are fairly rare now (1977). Their tendency to grow a big frame first and cover it with meat later make them a poor fit for today's conditions. The meat yield is rather disappointing until they are six months or more of age. No fowl with black plumage or dark or willow shanks has ever remained popular in this country for long. In the Giant's case, this is unfortunate because good specimens do have a different kind of appeal—mainly due to their size.

LANGSHANS

Class—Asiatic
Varieties—Black and white
Standard Weights:
    Cock—9½ pounds
    Hen—7½ pounds
    Cockerel—8 pounds
    Pullet—6½ pounds
Color of Skin—White
Color of Egg Shells—Brown
Use—A general purpose fowl for the production of meat and eggs. The general shape of the Langshan makes it better suited to roaster and capon use than to slaughter as fryers.
Characteristics—Langshans appear very tall. Long legs and tails carried at a high angle contribute to this illusion. Langshans are an active bird, and their long legs enable them to cover ground rapidly. Langshan females do go broody and usually make quite good mothers. Langshans' feet and legs are feathered but not nearly so fully as either Cochins or Brahmas.
Origins—Langshans originated in China and are considered one of our oldest breeds.
General Statement—Langshans enjoyed considerable popularity in the United States during the latter part of the 19th century. Today they are largely an exhibition fowl. The black variety is known for its deep greenish sheen when viewed in the proper light conditions. Many other breeds were created using Langshan blood in the foundation matings. Langshans are a good general utility breed, but their long legs and narrow body

conformation leave much to be desired as a meat bird by today's standards.

LEGHORNS

Class—Mediterranean
Varieties:
- Single Comb Dark Brown
- Single Comb Light Brown
- Rose Comb Dark Brown
- Rose Comb Light Brown
- Single Comb White
- Rose Comb White
- Single Comb Buff
- Single Comb Black
- Single Comb Silver
- Single Comb Red
- Single Comb Black Tailed Red
- Single Comb Columbian

Color of Skin— Yellow

Color of Egg Shell—White

Use—An egg-type chicken, Leghorns figured in the development of most of our modern egg-type strains.

Characteristics—A small, spritely bird with great style, the Leghorn has relatively large head furnishings (comb and wattles) and is noted for production of a large number of eggs. Leghorns rarely go broody.

Origin—Leghorns originally came from Italy. They take their name from the city of Leghorn where they are considered to have originated.

General Statement—A very active and quite noisy fowl, Leghorns like to move about. They are good foragers and can often glean much of their diet ranging over fields and barnyards. Leghorns are capable of considerable flight and often choose to roost in trees if given the opportunity. Leghorns and their descendants are the most numerous breed we have in America today.

MINORCAS

Class—Mediterranean
Varieties—Single Comb Black, Rose Comb Black, Single Comb White, Rose Comb White, Single Comb Buff

Standard Weights—Single Comb Black:
    Cock—9 pounds
    Hen—7½ pounds
    Cockerel—7½ pounds
    Pullet—6½ pounds
  All others:
    Cock—8 pounds
    Hen—6½ pounds
    Cockerel—6½ pounds
    Pullet—5½ pounds
Color of Skin—White
Color of Egg Shells—White

Use—Developed for the production of very large chalk-white eggs, the Minorca is today principally an exhibition fowl.

Origin—Developed in the Mediterranean area where they take their name from an island off the coast of Spain. Development may have been as an offshoot of the Spanish breed.

Characteristics—The largest of the Mediterranean breeds, they are long, angular birds that appear larger than they are. They have long tails and large wide feathers closely held to rather narrow bodies. Minorcas have relatively large combs and wattles.

General Statement— Good Minorcas are stately, impressive looking birds and can give a fair return in eggs, although in recent years they have not been intensively selected for that purpose. They are rather poor meat fowl because of their narrow angular bodies and grow slower than some other breeds. Minorcas rarely go broody, are very alert, and are fairly good foragers.

MODERN GAMES

Class—Game
Varieties:
    Black Breasted Red
    Brown Red
    Golden Duckwing
    Silver Duckwing
    Birchen
    Red Pyle
    Black
    White

Standard Weights:
>Cock—6 pounds
>Hen—4½ pounds
>Cockerel—5 pounds
>Pullet—4 pounds

Color of Skin—White
Color of Egg Shells—White or light tint
Use—A strictly ornamental fowl.
Origin—Modern Games were developed in Great Britain.
Characteristics—A very tightly-feathered bird with long legs and a long neck, giving it a very tall, slender appearance.
General Statement—The males of the Modern Games should have their combs and wattles removed in order to enhance their long, slim shape. The feathers of Modern Games should be short, hard, and held very close to their bodies. They do not stand cold weather well because of their shortness of feathers, and they need plenty of exercise to maintain muscle tone.

NEW HAMPSHIRE REDS

Class—American
Varieties—None
Standard Weights:
>Cock—8½ pounds
>Hen—6½ pounds
>Cockerel—7½ pounds
>Pullet—5½ pounds

Color of Skin—Yellow
Color of Egg Shells—Brown
Use—A dual-purpose or general-duty chicken that has been selected more for meat production than egg production. Medium-heavy in weight, it dresses a nice, plump carcass as either a broiler or a roaster.
Characteristics—Possesses a deep, broad body, grows its feathers very rapidly, quite prone to go broody, and makes a good mother. Most pin feathers are a reddish buff in color and, therefore, do not detract from the carcass appearance very much. The color is a medium to light red. The comb is single and medium to large in size; in the females it often laps over a bit.
Origin—New Hampshires are a relatively new breed having been admitted to the Standard in 1935. They represent a

specialized selection out of the Rhode Island Red breed. By intensive selection for rapid growth, fast feathering, early maturity, and vigor, a different breed gradually emerged. This took place in the New England states chiefly in Massachusetts and New Hampshire from which it takes its name.

General Statement—A good medium-sized meat chicken with fair egg-laying ability. Some strains lay eggs of dark brown shell color. New Hampshires are competitive and aggressive. The hen's feathers, being fairly light red, often fade considerably when the birds are allowed out in the sunshine. New Hampshires are the bird that were initially used in the Chicken of Tomorrow contests and, thus, led the way for the modern broiler industry.

OLD ENGLISH GAMES

Class—Game
Varieties—Black Breasted Red, Brown Red, Golden Duckwing, Silver Duckwing, Red Pyle, White, Black, Spangled
Standard Weights:
  Cock—5 pounds
  Hen—4 pounds
  Cockerel—4 pounds
  Pullet—3½ pounds
Color of Skin—White
Color of Egg Shells—White or light tint
Use—Old English Games are strictly an ornamental fowl.
Characteristics—A small, tightly feathered bird, Old English Games are very hardy, extremely active, and are very noisy.
Origin—Old English Games are the modern day descendants of the ancient fighting cocks. They are associated with England, but their heritage is almost worldwide. They have changed little in shape or appearance in more than 1,000 years.
General Statement—Old English have figured in the development of many other breeds. The mature cocks should be dubbed (have the comb and wattles removed) with a characteristic cut. This is in keeping with their heritage. Old English hens usually show broodiness, but they are so small and aggressive as well as defensive that they are not always the best choice as mothers. Old English are capable of considerable flight and capable of reverting to a feral state in some areas. They are the domestic breed most like the wild jungle fowl in appearance.

ORPINGTONS

Class—English
Varieties—Black, White, Buff, Blue
Standard Weights:
    Cock—10 pounds
    Hen—8 pounds
    Cockerel—8½ pounds
    Pullet—7 pounds
Color of Skin—White
Color of Egg Shell—Brown
Use—A heavy, dual-purpose fowl for the production of both meat and eggs. Orpingtons do exhibit broodiness and generally make good mothers.
Characteristics—Orpingtons are heavily feathered with feathers fitting rather loosely to the contours of their bodies. This gives them a massive appearance. They are quite docile and stand confinement very well. Also, because of their feathering, they endure cold temperatures better than some other breeds.
Origin—Orpingtons were developed in England at the town of Orpington in Kent County during the 1880s. They were brought to America in the 1890s and gained popularity very rapidly. Much of their popularity was based on their excellence as a meat bird. As the commercial broiler and roaster market developed, the Orpington lost out partly because of its white skin.
General Statement—Orpingtons are impressive birds and are a good, general utility fowl. Existing only in solid as opposed to patterned colors, they are equally at home on free range and relatively confined situations. Orpington chicks are not very aggressive and as a result are frequently the underdogs when several breeds are being brooded together.

PLYMOUTH ROCKS

Class—American
Varieties—Barred, White Buff, Partridge, Silver Penciled, Blue, Columbian
Standard Weights:
    Cock—9½ pounds
    Hen—7½ pounds
    Cockerel—8 pounds
    Pullet—6 pounds

Color of Skin—Yellow
Color of Egg Shells—Brown
Use—A dual-purpose or general duty, medium-heavy fowl for the production of meat and/or eggs.
Characteristics—Docile; normally will show broodiness; possesses a long, broad back; a moderately deep, full breast and a single comb of moderate size.
Origin—Developed in America in the mid to later part of the nineteenth century. The barred variety was developed first. It was noted for its meaty back and birds with barred feathers brought a premium on many markets. Most of the other varieties were developed from crosses containing some of the same ancestral background as the barred variety. Early in its development, the name Plymouth Rock implied a barred bird, but as more varieties were developed, it became the designation for the breed.
General Statement—Plymouth Rocks are a good general farm chicken. Some strains are quite good layers while others are principally bred for meat. White Plymouth Rock females are used to mother most of the commercial broilers produced today. Plymouth Rock hens usually make good mothers. Their feathers are fairly loosely held but not so long as to easily tangle. Generally, Plymouth Rocks are not extremely aggressive, but since they tame quite easily, some males and hens, too, are big and active enough to be quite a problem if they turn aggressive. Breeders should be aware of the standard weights and not select small or narrow birds for the breeding pen. The wide, straight back is a definite breed characteristic and should be maintained. Common faults include shallow breast, high tails, narrow bodies, and small size.

RHODE ISLAND REDS

Class—American
Varieties—Single Comb, Rose Comb
Standard Weights:
    Cock—8½ pounds   Cockerel—7½ pounds
    Hen—6½ pounds   Pullet—5½ pounds
Color of Skin—Yellow
Color of Egg Shells—Brown
Use—A dual-purpose or general-duty, medium-heavy fowl. Used more for egg production than meat production because of its

dark-colored pin feathers and its good rate of lay.

Characteristics—Possess a rectangular, relatively long body. Typical "Reds" have a very dark, red color and are good egg producers. Most Reds show broodiness, but this characteristics has been partially eliminated in some of the best egg production strains. The Rose Comb variety tends to be a bit smaller most of the time but actually should be the same size as the Single Combed variety. The red color tends to fade somewhat after prolonged exposure to the sun.

Origin—Developed in Massachusetts and Rhode Island, early flocks often had both single and rose combed individuals because of the influence of Malay blood. It was from the Malay that the Rhode Island Red got its deep color, strong constitution, and relatively hard feathers.

General Statement—Rhode Island Reds are a good choice for the small flock owner. Relatively hardy, they are probably the best egg layers of dual-purpose breeds. Reds handle marginal diets and poor housing conditions somewhat better than other breeds and still continue to produce eggs. They are one of the breeds where exhibition qualities and production ability can be quite successfully combined in a single strain. Some "Red" males may prove to be quite aggressive. Typically, Rhode Island Reds should be dark red in color. The use of medium red or brick red females in breeding pens should be avoided because this is not in keeping with the characteristics of the breed. Likewise, you should not breed from undersized individuals or birds that have black color throughout their body feathers. Remember, though, that black color in the main tail and wing feathers is normal and, therefore, a desired condition.

SUSSEX

Class—English
Varieties—Speckled, Red, Light
Standard Weights:
    Cock—9 pounds
    Hen—7 pounds
    Cockerel—7½ pounds
    Pullet—6 pounds
Color of Skin—White
Color of Egg Shell—Brown
Use—A general purpose breed for the production of meat and/or

eggs. One of the best of the dual-purpose chickens, a good all-around farm fowl.

Characteristics—Medium in size with a rectangular body. Sussex will go broody, make good mothers, and are alert. They are good foragers and were it not for their white skin could easily have become one of the leading fowls in America.

Origin—Sussex originated in the County of Sussex, England, where they were prized as a table fowl more than 100 years ago. They continue to be a popular fowl in Great Britain and the light variety has figured prominently in the development of many of their commercial strains. Sussex are one of the oldest breeds that are still with us today.

General Statement—Sussex are an alert and active bird. The speckled variety is especially attractive with its multicolored plumage. This breed is one of the best combinations of both exhibition and utility virtues to be found anywhere in the poultry world. They have never gained in the popularity in the United States that they have enjoyed in Canada, England, and elsewhere.

WYANDOTTES

Class—American

Varieties—White, Buff, Columbian, Golden Laced, Blue, Silver Laced, Silver Penciled, Partridge, Black

Standard Weights:
    Cock—8½ pounds
    Hen—6½ pounds
    Cockerel—7½ pounds
    Pullet—5½ pounds

Color of Skin—Yellow

Color of Egg Shells—Brown

Use—A dual-purpose or general duty fowl of medium weight; it is useful for the production of either meat or eggs.

Characteristics—Quite docile, will go broody, has a full, deep curvy body, a rose comb.

Origin—America. The Silver Laced variety was developed in New York State and the others in the north and northeastern states in the latter part of the nineteenth century and the early years of the present one.

General Statement—Wyandottes are a good fowl for small family flocks kept under rugged conditions. Their rose combs do not freeze as easily as single combs, and the hens make good mothers. Their attractive "curvy" shape, generally good disposition, and many attractive color patterns (varieties) make them a good choice for fanciers as well as farmers. Common faults include narrow backs, undersized individuals, and relatively poor hatches. Also, it is not uncommon to see single combed offspring come from rose combed parents. These single combed descendants of Wyandottes should not be kept as breeders.

GEESE

The domestic geese we have today are thought to be descendants of the gray-leg goose of Europe. We have, through centuries of domestication, varied the shape and size of some geese and selected for some color variations. Considering other animals though, the goose has perhaps been changed less than any of them by its association with man.

It is very much a bird with individualism and personality. Despite the fact that geese make phenomenal gains on very inexpensive diets, we haven't developed the mechanics of mass goose production the way we have with other fowl. Why? Well, we still may not be equal to "that dumb goose."

The goose is extremely alert, imprints easily, has few health problems, and, to date, has not overpopulated his part of the world. Let's look closely at the merits of geese and how to benefit from them. Geese can be raised in any part of this country. They do not require water for swimming, but they enjoy it. An ample supply of water for drinking should be available at all times.

Most geese are raised in small flocks on general farms and acreages (Fig. 6-1). There are few large, commercial growers. A market for geese exists, but it may be restricted in some areas due to the lack of processing plants equipped to handle geese.

Getting Started

Most people enter the goose business by purchasing day-old

Fig. 6-1. Geese are raised in small flocks.

goslings. Alternatives are to purchase adult breeding stock (usually a pair or a trio—one male and two females), purchase hatching eggs and set them under a broody chicken hen or in your own incubator, or to purchase started goslings.

Day-old goslings are quite expensive and always seem to be in short supply. Most hatcheries and the larger feed stores can order goslings for you, but you may have to place your order well in advance. This is preferred to ordering small numbers to be shipped direct to you from distant suppliers. Small shipments are subject to more stresses, including chilling en route.

Toulouse, Embden, African, Pilgrim and Buff are the more common breeds of meat-type geese with the first three listed being the larger. Chinese, Canadian, Egyptian, and Sebastopol are smaller and possess a unique quality admired by some people. All are good for home consumption. If the sale of live market geese is considered, the better choices would be Embden, Toulouse, or a cross between the two.

Diet

Geese are good foragers. They require little more than good grass pasture after they have been started.

They need good quality chicken starter in crumbolized or pelleted form. Because a few medications used with chickens may prove toxic to geese, it is safest to use a nonmedicated starter. Provide water they can easily reach for drinking but arranged so they cannot easily foul it. Water should be at least 1 inch deep.

Geese can be given small amounts of chopped grass or allowed to roam in a small enclosure after about seven days of age. Turning them out must be based on the weather conditions and the availability of grass. Letting them out on cold bare ground does little good. At one day old, geese chill quite easily, but by 10 to 14 days of age they show very little desire for the warmth of a brooder. Either the heat of a brooder or a mother hen is appreciated on occasion and nighttime roosting until they are quite large.

The female goose is a good mother. Chicken hens and various kinds of brooding devices can be used as foster mothers.

As the goslings grow, they will show a preference of grass and certain weeds over commercial mash. Mash should be available to them until they are completely feathered. Even though they may eat very little, it contains needed minerals and vitamins.

Generally, goslings prefer grasses to broad-leafed crops. This is the basis for their use in such crops as strawberries as "weeders." They also prefer new and tender growth. Therefore, it is often good policy to mow the taller, coarser growth from the goose pasture periodically in order to cause new growth to appear.

As the goose approaches maturity or with the onset of cooler weather in the fall, you should provide corn or other high-energy grain to promote finishing. Often geese appear to be quite well-feathered and nearly grown at about 14 weeks of age. At this time they can be slaughtered. If held beyond this time, they frequently proceed to grow another crop of feathers and will be hard to pick until these feathers mature (grow out of the quill or sheath), which may not occur until the onset of cool weather in the fall.

An acre of good pasture will support 20 to 30 geese, depending upon the season. It is good policy to avoid chemical treatment of the goose pasture or nearby areas during, and prior to, using it for goose grazing.

Adult geese being maintained as breeders or potential breeders should have access to a good poultry mash. They may not eat much of it but will require some, depending upon the conditions of their pasture. During the winter months, shelled corn and good alfalfa hay will be relished by them. In the summer, cannery waste is often fed to geese. Fresh vegetable trimmings and garden weeds will be sorted and largely consumed by geese.

The mature goose is a relatively light eater, considering the size of the bird, but it may be quite selective in its dietary choice. Make sure you are providing the things it will actually consume.

Housing

Geese are very hardy and stand all kinds of weather quite well as adults. A good brooder house with at least 1 square foot of floor space per bird is desirable. It should have a good dry floor, and this should be covered with 2-3 inches of absorbent litter, such as shavings or chopped straw. Heat lamps or a brooding hover with heat source will be needed. The temperature should be 85 to 90 degrees Fahrenheit at 3 inches above the litter the first few days, and after one week can be reduced five to seven degrees per week until room or outside temperature is reached. The brooder house should have the corners blocked or rounded to prevent crowding in them. Also, a draft shield or brooder guard, consisting of a strip of cardboard or tin, should encircle the brooder about 4 feet from its center to prevent floor drafts and to keep the goslings from wandering away from the heat before they learn where to go if they are cold.

During the growing period, it is desirable for geese to have access to a shed. This gives escape from the sun on hot days and affords a bit of shelter from hard driving rain and wind storms.

Social Life

Geese are long-lived creatures. Some may live 20 years or more and be productive throughout. The first season (from 12 to 24 months) many geese lay few eggs, and these may not be fertile. From two to seven or eight years are considered the prime seasons for breeding geese.

Ganders may become quite quarrelsome when their mates are nesting, as they are very protective. For that reason, very small children should not be permitted in the goose yard during the mating season.

Some geese tend to mate in pairs, but most of the larger breeds will usually mate satisfactorily at a ratio of three or four females per male. In small pens some males may not tolerate another male and will spend most of their time in disagreement.

Male and female geese should be placed together in the breeding pens at least one month before eggs are anticipated. Geese will normally begin to lay in late February or early March. Considerable variation often exists from goose to goose, but an average of 25 eggs per female is about normal for the larger breeds. Chinese are often the better egg layers, and 40 to as many as 60 eggs are not uncommon from them. The relatively small number of eggs and frequent low fertility accounts for the high price of goslings.

MUSCOVY DUCKS

The advantages to raising Muscovies, rather than other breeds of duck, are numerous. First, they are very easy keepers. During the winter, we feed a mixture of whole corn and commercial pellets. (They relish lettuce, spinach, celery, and beet tops—any greens—as a special treat). Once the snow has melted, the ducks show little interest in the grains. They prefer to spend their days searching for insects and grasses in the fields. We do make feed available to setting females (although they, too, will often prefer to gather their own, until the last few days before hatching) and to young, confined ducklings, who get a mix of cracked corn and pellets.

One question that always arises in regard to raising ducks is that of providing swimming water. (Naturally, drinking water must be available at all times, unless the ducks are confined at night and have no access to food at that time.) Although ducks enjoy a pond, it is not necessary for health or breeding purposes. We've raised ducks for many years, providing only drinking water, and have never experienced fertility problems. Swimming pools can prove disastrous, as many novices have learned. Young ducklings can swim from birth, but if they are unable to get out of the water due to a low level in the pool, they will drown. Be sure that any water available to ducklings is in a shallow container, as they will also climb into the drinking water.

In addition to being excellent sitters, Muscovies tend to be very protective of their young. They will perch on their houses or on fencing, keeping a watchful eye on the ducklings. Woe to the duck, animal, or human who gets so close as to be threatening.

Since Muscovies are one of the few domesticated ducks capable of flight, only a roofed pen will restrict them to a small area. Although we have a large section of the farm fenced off for the ducks, one would be more apt to find them on the front lawn, in the cow or goat pens, or, as we happily discovered one summer, devouring flies from the backs of our pigs.

Muscovies lay during the night or early morning; therefore, restricting them to a specific area also facilitates egg collection and observation of laying progress. We learned by hard experience that it's best to confine the young ducklings in separate, fine-mesh pens (within the large pen). This is further protection against their drowning in the adult drinking water or pool (if provided), against their straying into another duck's brood (the mother will be none too gentle about sending the young intruder on his way), and against their squeezing through the fencing and being alone in a world of horses' hooves and hunting cats. Muscovies tend to be quite hardy. Once their adult feathering has grown, they need no special provisions for winter weather.

One problem that can develop when too many ducks are locked up together is that of feather pulling. Aside from providing additional enclosures, we've solved the problem by identifying the offender and putting him into the freezer. In our breeding experience, we have found feather pulling to be a rare occurrence. Fighting will develop when too many males are kept for breeding, and one decides to assert his dominance. A ratio of six ducks to one drake is about right. The males are ready for the freezer by 14 weeks of age. This is the time to make final selections for future breeders. When the hatching season is over (September or October), we retain one older male and cull the others. (We keep about six males from the first hatch from which to choose breeding stock. This insures the drake's readiness for breeding the following spring.) We generally keep our best females for two years, as the older ducks are more likely to lay both earlier in the spring and later in the fall. Naturally, those with poor laying or setting practices also end up in the freezer. Some ducks will hatch out two broods a year. They and their offspring are given preference when selecting breeding stock.

The Muscovy is an excellent meat producer. Although 14 to 17 weeks is the ideal age for slaughtering, we annually process breeders and find no difference in flavor—only a tougher skin on the older ducks. There are numerous varieties of Muscovies, and the white is often recommended for eating as it produces a more attractive carcass.

We prefer wood shavings for bedding, although straw, leaves, etc., can be used. Although not really necessary, nesting boxes can be provided, This becomes more advantageous as the number of layers increases for it discourages multiple setters. The small duck owner will find that his ducks will happily make their own nests, either in the bedding or under the wood pile.

Muscovy eggs must incubate for 35 days. Beyond this, we always allow an extra week, as we may have made a calculating error. If there is still no hatching, we remove the mother and start breaking open individual eggs and examining their contents for reasons.

The most common cause of duckling mortality that we have observed is that of newly hatched ducklings becoming waterlogged when trapped in a deep watering bowl. Another common cause of duckling mortality is not drying out after hatching late in the day, then freezing in the cold night air. When we find a soggy duckling, we take it into the house overnight and put it under a light bulb or into an empty, lit aquarium tank. Ducklings must be kept warm, dry, and free from drafts.

Occasionally a duckling will injure its leg as a result of catching it on a root or being laid on by a careless mother. Again, we take the youngster inside, usually with another duckling for company, rig up the lamp for warmth, and after it has recovered (usually three or four days), return both to the mother. We've never had a problem with the female rejecting a duckling because it had been removed for a period of time.

An added endorsement for some is that Muscovies are quackless. Although not mute (they make a kind of hissing sound), they are much quieter than most breeds.

PIGEONS

The use of meat-type pigeons to produce a tasty item for the table is something that home food producers should not overlook. Pigeons can be satisfactorily raised in any part of the United States. The young bird at approximately 28 to 30 days of age, is known as *squab* and is a very tasty addition to the home food menu. Generally speaking, pigeons are kept for one of three reasons: food production, racing or sport, or for fancy or ornamental purposes. We'll concern ourselves here chiefly with those pigeons which can be satisfactorily and efficiently maintained for the production of meat.

For successful squab production, you need a large bird with the ability to produce several nests per year and to feed the young so they will be ready for slaughter at an early age. Generally, the best breeds for these purposes are Kings, Carneau, Mondain, Giant Homer, or Texan Pioneers. Most of these breeds will weigh from 25 to 30 ounces each as mature adults. Each of these breeds exists in several varieties or colors. From the standpoint of squab production, the color is not important except that those with the lighter colored pin feathers are easier to pick or dress. Therefore, light-colored birds are normally selected. Keep them confined so the parents spend most of their time feeding the young.

The goal of raising young squab is to keep the baby pigeon growing and feathering rapidly, being ready for slaughter at 28 to 30 days of age (Figs. 6-2 through 6-6). For the individual not familiar with pigeons, secure mated pairs from one to three years of age. It's not easy to distinguish the sex of pigeons and since they do pair for life (unless physically separated), it is a good idea to purchase mated pairs from an experienced pigeon breeder. Keep only mated

Fig. 6-2. Squabs at two days.

Fig. 6-3. Squabs at eight days.

adult pairs in the squabbing loft because unattached individuals create a disruptive influence on the mated pairs in the enclosure.

The male pigeon normally drives the female to the nest and is responsible for her attention to family duties. Normally the female sets on the nest at night. The male sets on the nest during daytime, and they reverse roles about the same time each day. Two eggs are laid per clutch, and they hatch 17 to 19 days later. Both of the parents participate in the feeding of the young.

Baby Gets Pigeon Milk

The first diet that the baby gets is pigeon milk. This is actually a secretion or fluffing off the interior wall of the crop and looks about

Fig. 6-4. Squabs at 14 days.

Fig. 6-5. Squabs at 28 days.

Fig. 6-6. At 30 days this squab is ready for market.

like thick cream. It has a high protein content and causes the young squab to grow at an unbelievable rate. After six to seven days, the flow of pigeon milk begins to subside. The parents begin transferring partially digested food particles from their crop to the youngsters, along with some remaining pigeon milk. By about 10 days of age, the pigeon milk is no longer a factor. The transfer is largely by mechanical operation wherein the parents pick up food particles, soak them for a short period of time in the crop, and regurgitate them into the beak of the young. This process continues throughout the period of time that the squab remains in the nest.

Ready in 28 Days

At approximately 28 days of age, the young pigeons will be heavier than their parents, and at this time they are ready for slaughter as squab. If they are not slaughtered at this time, they will leave the nest, begin to fly, learn to eat by themselves, lose some weight, and take on a firmness of flesh that is somewhat undesirable.

Good pairs of squabbing pigeons will normally lay eggs again at the time their current nest of youngsters is from 10 to 14 days of age. They will be incubating one pair of eggs and feeding a pair of squabs. Therefore, it becomes very important that the parents have easy access to an ample supply of feed because they're actually working very hard at maintaining their own body weight, incubating a nest of eggs, and keeping a pair of hungry youngsters full.

This routine of doubling up enables the parents to be hatching the second brood at about the same time the first nest of youngsters is ready to leave and fly on their own. For this reason, equip each brooding compartment with two nests for every adult pair so that there is no problem finding a place to set up the second nest while the first one is still occupied with the current youngsters.

Two satisfactory diets are available for the feeding of pigeons. One is a blend of mixed grains consisting of corn, wheat, kafir or milo, and Canadian or Maple peas. Blending this to about 15 percent protein gives a good overall diet for pigeons when combined with mineralized grit and a small amount of oyster shell. Another way, often used in commercial squab production, is to feed pellets containing approximately 15 percent protein, made from mainly vegetable sources. Though pellets are made from cheaper grains, they do incorporate mineral and vitamin supplements. Pellets are digested more rapidly than grains, and this is felt to be an aid in the rapid production of squabs. The disadvantage of pellets lies in the fact that the pigeons consume more water when eating pellets. This frequently results in a messier nest and a damp loft.

Loft Requirements

Frequently, 10 to 20 pairs of breeding pigeons are kept in a single loft. Allow 5 to 6 square feet of floor space plus a double nest for each of the pairs. There should be sufficient roosting perches throughout the pens so that each bird not actually occupying a nest can find an individual perch in order to avoid undue fighting and competition for favorite areas.

Pigeons generally prefer exposure to direct sunlight during part of every day and should be provided with a well-ventilated loft. Pigeons are remarkably healthy if given reasonable care, but they are quite sensitive to roosting in drafty situations or being confined in areas of high humidity. The major disease problem in the production of pigeons is *trichomoniasis,* commonly known as canker.

This disease can be controlled with any one of the several compounds that are on the market, but the owner must be constantly on the alert for the appearance of small, yellowish sores in the mouth and throat area of the young squab. Adult pigeons may be carriers of this malady without being visibly affected themselves. The young are frequently affected in the process of feeding. Lofts should be constructed to be absolutely free of rats and mice and to keep out wild birds. Wild birds not only bring in parasites, but they are also frequent visitors at the feed trough, which can be mighty expensive. Rats and mice may carry disease, frequently kill young squabs, and cause hens to desert the nest while incubating eggs.

Squabs are normally slaughtered and picked on a whole carcass basis and marketed either as individuals or by the dozen on ice pack boxes of fresh squabs.

It is desirable to keep accurate individual records of each breeding pair of pigeons in the squabbing loft, weighing the squabs at the time of removal and keeping track of the number produced. This will enable the operator to retain individual squabs from his best producing pairs for future breeders. The ability to reproduce at a satisfactory rate is a highly inheritable trait, so you want to be sure that replacement breeders come only from the most prolific parents.

If you're looking for something different in a home meat production, don't overlook pigeons and the production of squab. It is indeed a gourmet item for the backyard farmer.

TURKEYS

Turkeys are about as easy to raise and care for as any other poultry. It is mostly a matter of how you raise them and selecting the right breed for your location and space.

The most critical problem with turkeys, whether they were hatched from eggs under a hen or poults bought from a feed store, is getting them started feeding. With right attention the first week, they are usually slight trouble from then on.

Getting newly hatched poults to feed is sometimes more difficult than when poults are bought a week or two old. By then the laggards have been weeded out.

If the hatched poults refuse to eat at first, make sure they do drink clean water by dipping their beaks in it several times the second day after hatching. They really do not need to eat for the first 48 hours. Just dipping their beaks in the water and scattering trails of feed on clean white paper will usually encourage them to start feeding.

If these measures fail, you can get them started feeding by loading a small bread crumb with black pepper, rolling it into a pill about the size of buckshot, and stuffing it down them. One or two treatments always seems to do the trick.

Sometimes a weak poult will get pecked. Take that one out and feed it separately until it gains strength.

Selecting the right kind of turkeys contributes to the success of raising them under different climatic conditions. In the cold and harsher climate of Montana, we always raised the big broad-breasted bronze turkeys. The toms would weigh out more than 30 pounds after spring hatching. There were several reasons for this. The bronzes are long legged turkeys and they seem to withstand cold and wet weather better than other breeds. They are completely self-sufficient in 10 weeks and begin leaving the hen to fly up and roost in the trees or along the roof line of buildings where nothing can bother them. Although turkeys tend to frighten easily, it is only when they are confined in huge flocks that they really panic and injure themselves.

Since moving to Arizona, we have not kept any turkeys for breeding. Rather we buy one and two-week-old poults which are readily available at all the feed stores. These poults are generally in the same brooder with baby chicks in the feed store and have learned to eat from the feed hopper. When buying only a few poults, it is easy to watch them a few minutes and pick out the good feeders.

We now raise the smaller white Beltsville rather than the big bronze turkeys which are not suitable for our limited space, and find them a highly practical meat bird when raised under limited conditions. As white birds they stand the hot sun much better than the bronze breeds, which are actually almost black. They are table ready much younger, around 16 weeks for the hens, and are less excitable than the larger breeds. We also found that low fencing was sufficient to keep them penned.

To start the poults in this warm climate, we simply got a big cardboard box large enough for a dozen or two poults and covered the bottom with a thick layer of peat moss and hung a 100-watt red light bulb in one corner. After two or three weeks in this arrangement, depending on their age when bought, they have enough feathers to be turned out in the open where the first thing they do is run around with their wings stretched out trying to fly. After that they are pretty much on their own with fresh water and feed always before them.

In hotter climates, turkeys need plenty of shade (Fig. 6-7). Our citrus trees provided that along with low roosting limbs.

What we like most about turkeys is their remarkable ability to feed on insects. A turkey will have a grasshopper in a flash while a chicken is looking around wondering where it went.

One year in Arizona we raised turkeys from week-old poults in a fenced quarter-acre (3-foot woven wire fence with two strands of barb). On all sides the neighbor's ground was swarming with grasshoppers, and our turkeys almost lived on those high protein grasshoppers. With the grasshoppers and Johnson grass which we cut along the ditch banks, those turkeys grew out at little cost. Since

Fig. 6-7. Turkeys grown in the warmer sections of the country need open range and lots of shade.

turkeys are not much for scratching, they can be turned into garden areas which would be damaged by chickens to feed on insects.

Turkeys, with their strong beaks, will consume the discarded center stalks and roots of cabbage and lettuce. They thrive on green grasses of all types. In more open spaces they will range out and keep the growth completely free of grasshoppers and all other insects.

When turkeys are in the open, they like to roost up high. It is better if they do so, since nothing can bother them. If they are penned with chickens, arrange it so they can roost outside and they will be healthier.

Frequently young turkey hens will begin laying very early. This is especially true of the smaller breeds. Turkey eggs can be used just as hen eggs. If they are on open range, the turkey hens tend to be secretive and conceal their nests.

You do not have to wait until fall to enjoy your turkeys. At eight to 10 weeks of age, they make excellent heavy fryers, and many are used that way.

For just a few turkeys, a balanced formula of growing mash from the feed store is about the cheapest way to go. Turkey growing mash should be 25 to 28 percent protein, calcium fortified. If they are ranging where they can be partly self-sufficient, a good feed is a mixture of rolled oats with fresh or dried marigold blossoms broken and tossed in with the feed.

We always related turkey raising to the family food supply and as inexpensively raised poultry that can be marketed profitably. In both respects, turkeys have got to rank pretty high as a worthwhile addition to the economy of a small farmstead.

Turkeys have a remarkably high percentage of edible meat per pound of live weight—a dressing percentage of 80 percent in fact. A 12-pound turkey hen will yield 9 pounds 6 ounces of meat or 4.3545 kilograms. Turkey meat ranks with fish in protein value.

In confined feeding, 3 pounds or less of balanced feed stuffed into the turkey equals 1 pound live weight. Hardly any other meat source can equal that.

Turkeys do need a balanced ration that averages about 60 percent grain. If turkeys range in even a small area, their feed consumption is very low. After they gain their mature weight, it requires little feed to maintain them.

QUONSET HUT

You have little money and not much time, but still desire an effective; safe, and attractive addition to your farmstead—a poultry house that will provide maximum comfort and utility and look better than a marriage of old boards and chicken wire. The roof and walls are formed from a single piece of Homasote—the longer the sheet, the more floor space or height you can obtain. Homasote is a dense, soft gray material made from recycled newspapers and is waterproof. It is available at most any lumberyard and comes in ½-inch by 4 feet by 8 feet sheets for under $5. It can be cut with a Sheetrock knife. Do not confuse Homasote with Cellote, such as in ceiling tile, which is tan in color, not waterproof, and is inferior to Homasote. The ends of the coop are of ¾-inch exterior plywood or inch boards nailed to batten strips. Boards provide better nailing for the top, but plywood is quicker and easier to cut. The floor is also plywood (½-inch is fine) or boards nailed to 2 by 3 joists. See Fig. 6-8.

The large back door is for cleanout and egg access, and the small front door is for the use of the birds. The aluminum drip caps are optional. The entire unit is set up on concrete blocks and surrounded by a wire fence enclosing a yard of suitable size.

If the sleeping perch is put near and parallel to the cleanout door, most of the droppings can be easily removed. Once the birds are accustomed to their new house, they may be shut up in it nightly to keep them safe from predators and released as your first animal chore in the morning.

The doors should be replaced by screen doors in warm weather to avoid overheating and suffocation, should you neglect to release the birds early in the morning, and to provide ventilation at all times. Six mature birds will be comfortable at night in all but the most severe winter weather if bedded with ample hay.

If using an 8-foot piece of Homasote, use the formula $C = pi(d)$, with the following substitution: $96 \times .2 = 3.14d$, where 96 is the Homasote length in inches, multiplied by 2 to get the total circumference (we're only building half the circumference—3.14 is an approximation of pi, and d is the diameter or width of the floor). Thus, the floor width should be about 61 inches. We used a 5-foot width, thus cutting a standard 10-foot board into usable lengths with no waste. Ten-foot Homasote translates as follows: $240 = 3.14d$ or $d = 76$ inches.

Draw the base width to scale on a piece of paper, and with a compass, construct an arc joining the endpoints. The perpendicular line joining the midpoint of the line to the top of the arc will give you the height (in scale) of the building. For a taller building (as for

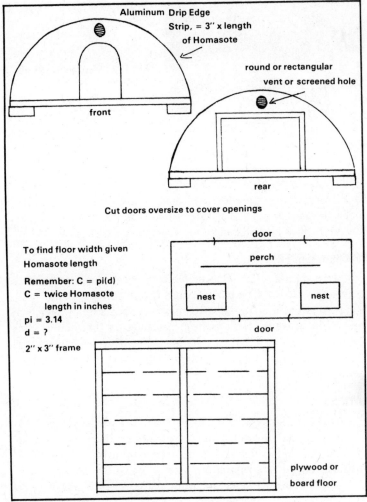

Fig. 6-8. Construction details for the quonset hut.

geese), make the floor narrower. For more floor space (chicks or quail), make it wider.

Assemble the floor, nail the Homasote to one side, and bend it over to the other side. Adjust to get nearly an exact curve by binding about with rope or by pushing on one side while an assistant traces the curve on a sheet of plywood or preassembled end piece. Cut the end piece with a saber saw and nail it in place with 1¼-inch or longer roofing nails. A 3-inch strip of aluminum flashing will add to the life

Fig. 6-9. Araucanas admire their new home.

of the unit. Apply as many coats of paint as you care to and install ventilators, doors, perches, and nests. Now you have a quick, cheap, attractive, and useful coop primarily for small flocks or trios (Fig. 6-9). It is also adaptable for dogs and cats.

CHICKEN FEEDER

Here are plans for a chicken feeder you can make out of scrap lumber (Fig. 6-10). The plans feature two important points—a lip along the top edge to prevent waste of feed, and a roller device on the top to prevent chickens from perching on the feeder, thus allowing droppings to fall into the feed. The plans are intended as general guidelines and may be adapted according to the size of the feeder needed and the materials at hand.

RATIONS

Poultry rations are normally designated according to the protein they contain. Several other components are also important, but when commonly fed ingredients are used to bring the protein to the proper level, the other items are usually not far off. This is understandable because we select feeds for chickens in the first place on the basis of what chickens eat and respond to with acceptable growth and/or egg production. So, let's look at some common poultry feeds and why they are used (Table 6-1).

Remember, the chicken needs a blend of energy, protein minerals, and vitamins (Fig. 6-11). Generally speaking, grains are fed for energy with sufficient animal by-products or milled feeds added to bring up the protein level. Small quantities of minerals and vitamins are also added to make up for any remaining deficiencies.

Let's take an example. You need a laying ration and have corn on hand and plan to buy a concentrate. The local feed store has a 42 percent (protein) concentrate, so how much of each do you need?

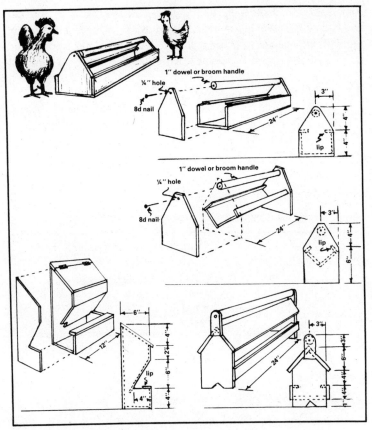

Fig. 6-10. Chicken feeder plans.

According to standard value tables, corn normally contains about 9 percent protein on an "as fed" basis. Then, if you use 300 pounds of corn and 100 pounds of concentrate:

300 pounds of corn at nine percent protein	= 27 pounds protein
100 pounds of concentrate at 42 percent protein	= 42 pounds protein
400 pounds of feed	= 69 pounds protein

This combination provides a 17.25 percent protein feed.

Now back to the hen. The above mix is adequate from a protein standpoint. It should also be okay for minerals and vitamins because the concentrate was formulated with the idea of being blended with grain. This mix would be quite high in energy, which could cause some hens to start picking. Therefore, you may want to consider some oats. Oats have more fiber and less energy than corn. Hence, the hen will eat a bit more oats in order to meet her energy requirements and in so doing will be fuller and feel more contented (and less likely to pick). Oats have a protein level closer to corn that to the concentrate so let's try:

275 pounds of corn at nine percent protein	= 24.45* pounds protein
25 pounds of oats at 12 percent	= 3* pounds protein
100 pounds of concentrate at 42 percent protein	= 42 pounds protein
400 pounds of feed	= 69.45 pounds protein

This combination provides a 17.36 percent protein feed.

Table 6-1. Common Poultry Feeds and Why They Are Used.

Items	Common Ones in Poultry Feeding	Principle Contribution
Grains	corn, millet, oats, milo, wheat, rice, barley	Energy; some vitamins and some protein*
Animal By-products	meat & bone meal, fish meal, meat meal, dried milk, dried whey	Protein; minerals and vitamins
Milled Feeds	wheat middlings, soybean oil meal, alfalfa meal, corn gluten meal, cotton seed meal, bran	Protein; minerals and vitamins
Minerals	salt, oyster shells, defluorinated rock phosphate, ground limestone	Minerals
Oils and Fats	corn oil, animal fat	Energy
Green Feeds (including fruits, vegetables & root crops	grass, sprouted grains, lettuce, comfrey, mangels, carrots, turnips, cabbage	Vitamins and appetite stimulants

Fig. 6-11. Chickens need energy, protein, minerals, and vitamins.

*Grains, though relatively low in protein, contribute considerable total protein in the ration because of the large amount used.

That should do the job protein-wise and still leave the hen contented. Notice, however, that protein is now over 17 percent, so if you had a lot of hens you might want to consider the economics of using:

300 pounds of corn at nine percent protein	= 27* pounds protein
25 pounds of oats at 12 percent	= 3* pounds protein
100 pounds of concentrate at 42 percent protein	= 42 pounds protein
425 pounds of feed	72 pounds protein

Using this mix, you get a 16.9 percent protein content.

If you place a value of 5¢ per pound on corn, 6¢ per pound on oats, and 12¢ per pound on concentrate, you will get mixed prices of 6.75¢ per pound for the first mixture, 6.81¢ per pound for the second, and 6.73¢ per pound for the third. These are small differences if you had 10 hens, but represent a lot of money to the commercial producer who has 200,000 layers.

A good ration for the person who wants to produce eggs and perhaps hatch some of them would be the following (Table 6-2).

In order to do an adequate job of completely formulating a ration, you need to have a table of feed ingredients that shows protein percentage, vitamin and mineral content, and the amino acid content of the protein. You should also have a table that shows the requirements of the chicken for protein, vitamins and minerals. These tables can be found in such books as *Poultry Production* by Card and Nesheim or the published tables of requirements as determined by the National Research Council. Numerous feed companies and ingredient suppliers also publish these tables.

If you use concentrates plus the usual grains, this is taken care of by simply working out mathematically the desired protein content. If you want to use uncommon ingredients or unusual mixtures, you must refer to the tables. The easiest procedure, in any case, is to calculate the protein desired and then check the other features for adequacy. Working the items of greatest volume first speeds up the procedure. For most chicken diets this will be corn (50 to 65 percent), soybean oil meal (20 to 30 percent), and the balance in by-product items and supplements.

There is nothing magic about feeding chickens or formulating their diets, but you must recognize that the chicken has specific needs. Failure to meet these needs can only result in disappoint-

Table 6-2. A Good Egg-Producing Ration.

Item	Amount in Pounds	Percent Protein	Pounds of Protein
Corn	1403	9	126.27
Soybean oil meal	300	44	132
Meat and bone meal	50	60	30
Fish meal	50	60	30
Alfalfa leaf meal	50	17	8.5
Calcium supplement	115		
Dikal	22		
Salt	5		
Vitamin premix	5		
Total	2000	16.34	326.77

ment to you. It will show up in reduced egg production, unsatisfactory growth, and poor health.

Take the laying hen as an example. She usually consumes feeds to meet her energy requirements. It is up to us to supply the other things she needs in the amount of feed consumed. This is the real reason for thoroughly mixing her food.

What about the hen running outside picking up bugs? Our reply has to be simply "Did you ever analyze a bug? You'll find it is very high in animal protein and can balance several weed seeds to make a pretty usable ration. This, of course, presupposed that there are plenty of clean bugs and well developed weed seeds.

The hen uses her food to maintain her body, manufacture the eggs she lays, maintain body temperature, and support movement. It is up to you to make sure she has all of the nutrients needed to do these jobs. If not, it's your fault.

HATCHING EGGS

Hatching eggs from your own hens can be a source of fun and education for the entire family (Fig. 6-12). Here are some basic rules to guide you.

You can tell a hen that's ready to brood if she stays on the nest after she lays an egg and defends the nest when another bird or a person comes near. She will often raise her feathers and make a raspy sound. When she stays on the nest overnight, you can allow her to incubate a nest of eggs for you.

The hen will need a nest of a comfortable size, which has a closable front and can be located where other birds or animals cannot annoy her. She should be provided with access to water and feed daily and allowed to move around for a few minutes.

Dust her thoroughly with a good insect powder before allowing her to "set." The number of eggs she can handle depends upon her size and the tightness of her feathers. Early in the season, she must be able to cover all of the eggs at all times.

The hen should be disturbed as little as possible. After the fourth day, the eggs may be candled to see whether or not they contain developing embryos.

To candle, hold the egg before a shielded bright light in a relatively dark room. Those that appear to have a reddish spiderlike body inside at four to seven days of age are fertile and developing. Those that appear to have a black loop at seven days are called blood rings—embryos that started to develop but have since died. Infertile eggs will look about the same after seven days as they did when

Fig. 6-12. Hatching eggs from hens can be fun and educational (photo by Larry Dunlap).

laid; only a little more yolk shadow and a little increase in the air cell size will be evident.

The broody hen is able to stay quiet for long periods during incubation because she is in a state of partial hibernation. Her temperature drops slightly and her metabolic processes slow somewhat. Thus, she can "set" for long periods without getting cramped.

When the first chicks hatch, it is well to remove them to a 95 degrees Fahrenheit brooding area until the remainder hatch. Otherwise, the hen may take the first hatched and desert the nest, allowing the other eggs to chill. On the twenty-second day, take the hen from the nest to a relatively secluded brooding coop and place her chicks under her.

Because of limited space and the possibility of disease being passed from the hen to the chicks, many people prefer to use an incubator, which really does only what the broody hen does (Table 6-3). You can build one yourself.

You'll probably use a still air incubator, in which the temperature at the embryo level is about 100 degrees Fahrenheit, the amount of warmth which the hen would give it. Other important factors are humidity, ventilation, and turning the eggs.

Humidity

Eggs developing embryos must dehydrate about 25 percent of the original contents during incubation. If they dry up too much or not enough, the chick's ability to emerge from the shell is adversely affected. Relative humidity should be held at about 60 percent and increased, to about 70 percent during hatching—the last three days. It then should be dropped just before the chicks are removed so they can fluff out nicely and more easily adapt to the cooler world outside.

Pans of water evaporating in the bottom of the incubator usually provide the humidity. Sponges may be added to increase it, and just prior to hatching a loosely woven cloth, damped with warm water may be used to line the bottom. This will also serve to help you easily remove egg shells. Opening and closing of vents will also help control the humidity.

Ventilation

The developing embryo is a living creature, so it needs to breathe oxygen and in turn give off carbon dioxide. The oxygen is taken from the air by diffusion through the shell if sufficient fresh air is available in the incubator.

The ventilators should be nearly closed when the eggs are first placed in it and opened gradually as incubation proceeds. When incubating less than a full set of eggs, adjust the vents for a proportionally smaller opening. Opening them too much makes it difficult to maintain the proper humidity and temperature.

Location of Incubator

The incubator should be placed in a draft-free location out of the direct rays of the sun. An ideal would be in a 70 to 75 degrees

Table 6-3. Incubation Period and Incubator Operation for Eggs of Domestic Birds

Requirements	Chicken and Bantam	Turkey	Duck	Muscovy Duck	Goose	Guinea	Coturnix Quail
Incubation Period (days)	21	28	28	35-37	28-34	28	17
Forced-Air Operating Temperature[1] (degrees F., dry bulb)	99¾	99¼	99½	99½	99¼	99¾	99¾
Humidity (degrees F., wet bulb)	85-87	83-85	84-86	84-86	86-88	83-85	84-86
Do Not Turn Eggs After	19th day	25th day	25th day	31st day	25th day	25th day	15th day
Operating Temp. During Last 3 Days Incubation (degrees F., dry bulb)	99	98½	98¾	98¾	98½		99
Humidity During Last 3 Days Incubation (degrees F., wet bulb)	90-94	90-94	90-94	90-94	90-94	90-94	90-94
Open Ventilation Holes One-Fourth	10th day	14th day	12th day	15th day	1st day	14th day	8th day
Open Ventilation Holes Further if Needed to Control Temperature	18th day	25th day	25th day	30th day	25th day	24th day	14th day

[1] For still-air incubators add 2-3° F. to the recommended operating temperatures.

Better hatchability may be obtained if goose eggs are sprinkled with warm water or dipped in lukewarm water for half a minute each day during the last half of the incubation period.

Fahrenheit area. Excessive variations of temperature would be harmful in incubation.

Care of Eggs before Incubating

Eggs should not be more than seven days old prior to being set. Beyond that point, hatchability declines. They should be kept away from sharp variations in temperature. A good location is one around 50 degrees Fahrenheit with a relative humidity of 70 to 80 percent. We use a shelf in the root cellar, although the crisper in the refrigerator would do as well. They should be turned daily. Don't wash them; use only clean, unwashed eggs.

Turning

Turning the eggs help prevent the embryo from adhering to the shell and also gives them some exercise as they reorient themselves. The hen turns the eggs every few minutes when she is setting, but with an incubator, two or three times a day will do.

The eggs can be marked with an X on one side and an O on the other, so you know which ones you have moved. The eggs will cool slightly while being turned but this shouldn't harm the process. In fact, it's recommended.

It is best if the eggs are turned an odd number of times each day so that the side that is up longest is staggered from day to day. Otherwise, the egg will be in the same position every night. When you turn the eggs, move them to a different part of the incubator to offset temperature variations. Continue to turn them from the first through the nineteenth days, but do not turn them after the nineteenth.

Reasons for Poor Hatches

If your hatch is poor, it's probably because of one or more of the following reasons: infertile eggs; eggs too old when set; parent stock weak, unhealthy, or fed a nutritionally deficient diet; improper care of eggs prior to incubation; shell contamination; eggs not turned often enough; temperature too high, too low, or too variable during incubation; too little humidity or occasionally too much humidity; improper ventilation; or oxygen starvation.

BUCKET INCUBATOR

An incubator needs to provide heat and a way of regulating it, as well as humidity and ventilation. Bearing these facts in mind, we set to work constructing an incubator that would work as well as the commercial ones and cost less. It cost even less than we expected—under $5.

We used a 3½-gallon plastic bucket with a tightly fitting lid that had originally held a photographic chemical, but happened to be idle at that time (Fig. 6-13). A box, washtub, or almost any kind of suitable container can be used. Ours proved to be somewhat small, especially when it came to hatching goose eggs, as it would only hold 15 chicken eggs or six to seven goose eggs.

Having selected a container, it is important to insulate it as well as possible to prevent heat loss and help keep the temperature stable inside the incubator. It should be totally enclosed except for air holes. Our bucket was not very thick-walled, so we lined it with aluminum foil to insulate it and reflect that.

We bought an automatic temperature regulator—available from Sears and farm stores such as Countryside. This was actually the only part of the incubator that we had to buy, apart from the light bulbs. If we had been very ingenious, we could have adapted a thermostat from some old household heating device. Temperature control is very important in an incubator, though, so it was probably worth buying one. We mounted the temperature regulator on the lid and wired it to the two 30-watt light bulbs that were to be the heat source. We used two light bulbs so that the eggs would not be deprived of heat should one of them burn out in the middle of the night.

We drilled three small holes in the side of the bucket about 1 inch from the bottom and made an observation hole in the lid about

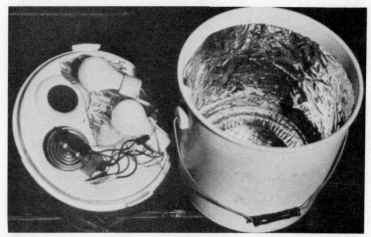

Fig. 6-13. A do-it-yourself bucket incubator (photo by P.J. Weeks).

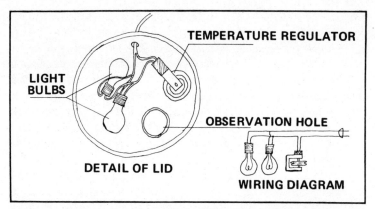

Fig. 6-14. Lid detail and wiring diagram (drawing by P.J. Weeks).

2½ inches in diameter (Fig. 6-14). These holes were to allow proper ventilation and circulation of heat in the incubator.

We placed an aluminum pie pan in the bottom of the bucket to hold water and maintain the correct humidity (Fig. 6-15). We made a grid out of ½-inch mesh wire to place over the pie pan and hold the eggs. A thermometer was attached so we could check the temperature—which should be at about 103 degrees Fahrenheit. We used a long chemical thermometer and suspended it from a wire through the observation hole so the bulb was about level with (but not touching) the tops of the eggs. You can buy a special incubator thermometer on a metal stand which sits on the grid with the eggs, but ours had the advantage that we could read the temperature easily without lifting the lid (and another advantage that we did not have to go out and buy it).

We did not use a hygrometer in our incubator, but it is nice to have one so you can keep a check on the humidity, which should be between 70 percent and 80 percent. We just kept the water pan filled and hoped for the best, which turned out to be good enough.

The temperature regulator can be adjusted by turning the knob until you get the correct temperature in the incubator. We found that it took some time for the temperature to stabilize, so we usually turned the incubator on the day before we put the eggs in. Putting the eggs in and taking off the lid to turn them will cause the temperature to drop considerably, but it will soon return to the set temperature. The regulator may need further adjustment during the incubation period, but don't be in too much of a hurry to adjust it.

The eggs must be turned throughout incubation. We turned ours three times a day, every eight hours, as this fitted into our schedule conveniently. For some reason, eggs are supposed to be turned on odd number of times—three, five, seven, or nine times a day. We marked each egg with a penciled X so we could easily keep track of which eggs we had turned. We also found the eraser end of a pencil a useful tool for turning the eggs. We made a note on the calendar of the day we started the eggs. That way we knew when to expect them to hatch. Chicken eggs take 21 days, and by the twentieth day you can hear cheeping sounds from the eggs. The eggs can be candled after about 14 days to see if there is an embryo developing inside.

We were somewhat pessimistic about our incubator in the beginning, but were encouraged when eight chicks hatched out of our first 14 eggs and 10 out of the second batch (Fig. 6-16). We also hatched duck and goose eggs and had a fair measure of success with them, and later some guinea fowl eggs. The guinea hen probably would have been happy to set on the eggs, but a varmint was stealing her eggs as fast as she laid them, so we decided to come to her rescue.

Our incubator was put together mainly from materials we had lying around the place. The electrical wire and fittings for the light bulbs were taken from an old table lamp we weren't using, the wire for the grid was left over from a cage we had made, and most people have a pie pan and aluminum foil in their kitchens. We bought the temperature regulator and the light bulbs which came to about $4.75. Even if you had to buy your container, electrical fittings, and thermometer, this incubator should cost you less than $10.

The materials required are as follows: a bucket or other available container, with tightly fitting lid; aluminum pie pan; a small piece of ¼ to ½-inch mesh wire, aluminum foil; two 30-watt light

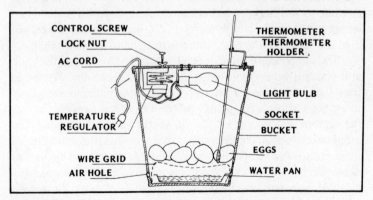

Fig. 6-15. Cross section of the incubator (drawing by P.J. Weeks).

Fig. 6-16. Healthy chicks are on their feet soon after hatching (photo by P.J. Weeks).

bulbs; electrical wire; plug; and light bulb sockets; automatic electric temperature regulator; and thermometer.

Too many or too few cockerels in the flock can affect the fertility of the eggs. More than one cock per 10 hens is too many, and less than one per 30 is too few. Outdoor temperature can also affect egg fertility—don't start collecting eggs too early in the year.

Choose medium-sized eggs for hatching. Do not save an egg with even the tiniest crack in it. Discard odd-shaped or thin-shelled eggs. We heard that one could tell the cocks from the hens by the shape of the egg. The rounder eggs are said to contain hens and the more pointed ones cocks, but we did not find this very reliable.

Eggs should not be kept in the refrigerator prior to incubation. Rather they should be stored in a cool place—55 degrees to 60 degrees Fahrenheit—in an open container where air can circulate around them. They should not be kept for too long either. We had the greatest success with eggs collected over a seven-day period. After 10 days the probability of their hatching declines rapidly.

Chicks hatch quite quickly under normal conditions, and although it may seem that they are taking too long, the temptation to help them out should be resisted as premature separation from the membrane can cause bleeding which is sometimes fatal to the chick. If a chick takes forever to get out of his shell, he is probably deformed in some way. The rate of abnormality in chicks seems to be quite high—we had two out of 25.

Chapter 7

Rabbits

Do rabbits have a place on a "real" farm? These small and docile creatures have long been considered the province of youngsters and, recent years, homesteaders, who raise food for their own use but not on a commercial scale. The French, Germans, and Italians and certain other nationalities eat a great deal of rabbit and, during hard times, Americans have put a good many on their tables, too. Generally, though, this little beast has been looked down upon in the United States.

Perhaps it's time for another look, because the much maligned rabbit has a great deal to offer anyone with an interest in livestock. The rabbit has much to offer anyone interested in good eating, and it certainly has merit for anyone who is interested in increased self-reliance (Fig. 7-1).

BREEDING STOCK

When you are raising rabbits for homestead meat production, you need breeding stock that's really productive. Some rabbits aren't good producers. The does don't have large litters, or they don't produce enough milk to grow the young rabbits quickly. Breeding problems (misses) are more common with poor stock.

There is no sense wasting your time, cage space, and feed money on poor rabbits. Get good stock in the first place.

Pedigrees do not guarantee good production. Fancy show rabbits often have not been culled and selected for efficient meat production, although some do quite well. Listen to what the seller

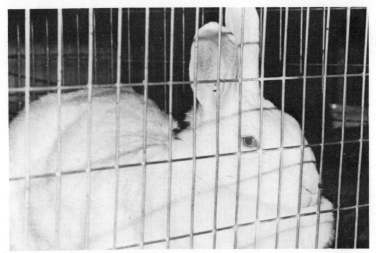
Fig. 7-1. Most farmers should raise rabbits for delicious, inexpensive meat.

has to say about his rabbits. If he has been keeping records and culls for stock that turns out large and fast-growing litters, these are the rabbits you want.

New Zealand Whites are the most popular commercial breed for meat production, but other breeds can do quite well. The commercial raiser gets a bonus for the white pelt. The processing plant may pay more for white fryers (they are faster to clean). The laboratory market, which pays very well, often demands a white rabbit.

For homestead production you might prefer a colored breed, simply because they are more attractive or interesting. There are more than 30 rabbit breeds to choose from. They range in size from the very small Brintanna Petites and Netherland Dwarfs—less than 3 pounds at maturity—to the giant breeds—the Flemish Giants and Checkered Giants and Giant Chinchillas, which sometimes weigh nearly 20 pounds as adults.

Meat rabbits are normally butchered as fryers at eight to 10 weeks of age. The slow-growing giant breeds are not necessarily larger at that age.

You really don't need purebred rabbits. "Just plain rabbits" can be good producers if the person who is raising them has paid serious attention to culling for his breeding stock. Once in a while you will find someone who is raising very nice rabbits that are common crossbred mixtures, but it is more likely that the serious rabbit producer will be raising a pure breed.

The smaller breeds might be a good choice for meat production for some folks. Polish weigh about 2½ pounds when mature. Only the Dwarfs and Petites are smaller. They don't take much room for housing, and people can raise them in a basement or the garage. They dress out to 1½ to 2 pounds, which is just about all the very filling rabbit meat an older couple or a single person might want for a meal.

In his very good beginner's book *Raising Rabbits the Modern Way,* Bob Bennett has praise for the smaller breeds that weigh 4 to 6 pounds at maturity, such as his favorite Tans. Dressed out at 10 to 12 weeks of age when they weigh 3 to 3½ pounds, they deliver almost as much meat as an eight-week-old New Zealand because there is less waste, he insists. Bennet feels the small breeds can produce meat more economically per pound, if you don't count the labor of keeping them a few weeks longer, and if you can find a really productive strain to buy.

Some people find 10-pound adult New Zealands a bit heavy to handle. For them, a smaller breed is the best choice.

Don't confuse weight with smallness. You want a rabbit that is built square and blocky with a lot of muscle meat. You don't want a 5-pound adult rabbit that's long and lanky, mostly bone, hardly enough flesh to make it worthwhile to dress the animal out, and that takes forever to grow. That's just a poor rabbit. Prices for good commercial New Zealand breeding stock usually range from $8 to $18 per rabbit, depending on age.

Culling is a way to improve a herd over the years, although it is slow and you're wise to simply buy better animals in the first place. Demand production from your rabbits. Get rid of does that produce small litters. You want at least six weaned fryers per doe, and some breeders insist on seven or eight. Get rid of does that produce undersized, lightweight bunnies at weaning. They are probably poor milk producers.

Keep breeding stock only from the healthiest, fastest-growing strains. The buck should be from a very good producing dam.

You should get at least 55 percent dress-out on a young fryer and a lot of meat on the bones. If your rabbits aren't doing that well and they have plenty of feed, you need better rabbits.

WOOD AND WIRE HUTCHES

Some backyard breeders are quite satisfied with wooden hutches, but they do present problems. Wooden hutches have to be cleaned more frequently than wire cages, they're harder to clean, and they're very much harder to keep sanitary (Figs. 7-2 and 7-3). Wood gets damp, rots out, and collects disease organisms. If the wood isn't protected by wire, the rabbits will chew it to bits, but if it is protected, manure is almost impossible to get out from between the wood and the wire.

With wire cages, the droppings and urine pass through. You can sanitize a wire cage quickly and thoroughly by running a blowtorch over the wire.

In most cases, a wire cage will cost you less than a wooden hutch even if you use recycled wood. If nothing else, the all-wire cage will last far longer. You can buy ready-to-assemble wire cages for $12-$15 from many rabbitry supply houses.

If you prefer to make your own rabbit housing, here are a few tips. If you decide to use wood, be sure there are no exposed frame surfaces the rabbits can chew on. They can go through a good-sized piece of wood in nothing flat. Remember that you'll have to clean the cage daily: uneven boards or an edge at the door, or a too-small door can cause real headaches. Do not use chicken wire or poultry netting, because the rabbits will pull at it with their teeth and dogs can tear it easily.

Fig. 7-2. Wooden hutches can be attractive and serviceable. They are usually more expensive than all-wire cages. The poultry netting offers no protection from predators, and the rabbits themselves can be injured by it. They do not have the working lifespan that wire cages have, and they are much more difficult to clean and maintain.

Fig. 7-3. The all-wire cage is more economical, much more sanitary, and nearly maintenance-free. The cage is simply hung from the rafters of a building. One drawback may be the difficulty of obtaining wire in the small quantities a homesteader needs. It's often cheaper to purchase the cage.

It's easy to make a wire cage. Doe cages are normally 36 by 30 inches by 18 inches high. Use 1 by 2-inch 14-gauge galvanized after welding mesh for the top and sides and ½ by 1 16 gauge for the floors. The "galvanize after welding" is important. The smaller spacing of the floor wire prevents babies' feet from going through, and the heavier gauge adds stability. Be sure the smooth side of the wire is up.

For a cage like this, you would need one piece 36 by 30 inches for the floor, one the same size for the ceiling, two 36 by 18 for the front and back, and two 30 by 18 for the ends. You'll also need a piece about 13 by 15 for the door. This wire is also available from rabbitry suppliers, although usually only in quantity.

Small metal fasteners—either J-clips or C-clips—can be used to fasten the pieces of wire together. A special pliers is available for either, although for one cage a regular pliers will work. (You could even dispense with the clips and just use short pieces of wire, wrapping them tightly around the sections to be joined together.)

Cut a hole for the door. This should be at least 12 by 14 inches, large enough to admit a nestbox. Carefully bend the cut wires back on themselves so neither you nor a rabbit can get snagged.

Do the same with the piece you cut for the door. (This is why you can't just use the piece you cut out; by bending back the cut wires, your door will end up about an inch short all around.)

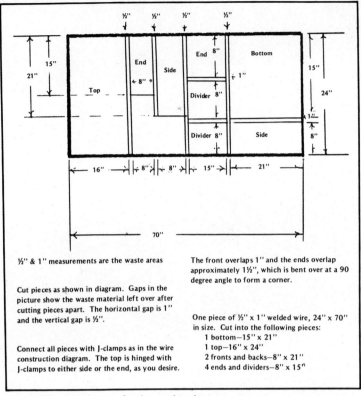

Fig. 7-4. Cutting diagram for the carrier pieces.

Some people like their doors to swing in, some out, up or down, and some even to one side or the other. Whichever you choose, "hinge" the door to the cage using the same clips.

Special door fasteners are available, or you could fashion your own out of a piece of wire or metal. There are no rules here. Most commercial cages have a length of rod beneath the center of the floor to provide added support. Hang your new cage at a convenient height in your barn, garage, or similar location by means of wires from the cage top to the building rafters.

The basic materials for this carrier consist of 15 feet of 1 by 2s and a piece of ½ by 1-inch welded wire, 24 by 70 inches. To start, cut the pieces, as shown in Fig. 7-4 and build with C- or J-clamps. The narrow spaces in Fig. 7-4 are waste areas that occur when the wire is cut, leaving smooth edges on all sides.

Once this is done, you can start construction of the framework (Fig. 7-5). This frame provides a sturdy surface on which to anchor

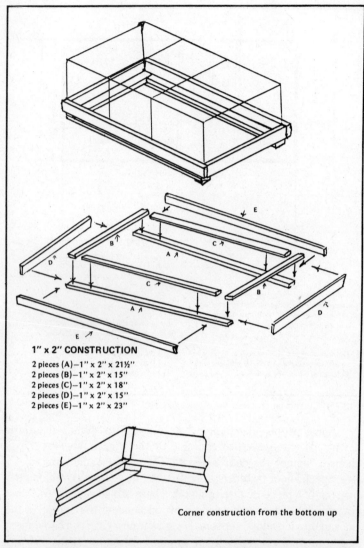

1" x 2" CONSTRUCTION

2 pieces (A)—1" x 2" x 21½"
2 pieces (B)—1" x 2" x 15"
2 pieces (C)—1" x 2" x 18"
2 pieces (D)—1" x 2" x 15"
2 pieces (E)—1" x 2" x 23"

Corner construction from the bottom up

Fig. 7-5. The wire housing rests on pieces A, B, and C. The unit is anchored to these pieces after the frame has been constructed. Pieces A and D are installed after A, B, and C are built and the housing is attached.

the wire unit and will minimize shifting of the wire section. It also provides a built-in handle (the gap between the floor and piece B as piece A rests on the floor). We also attached a piece of scrap 1 by 2 to the lid to provide a space to print your name or some form of advertising. Spray paint the 1 by 2 frame after the entire unit is

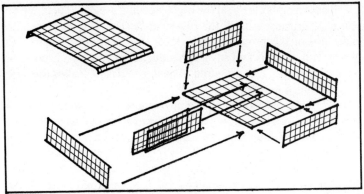

Fig. 7-6. Top wire construction for the carrier.

assembled. You can use C- or J-clamps to construct the wire unit and for the hinges (Fig. 7-6). A spring anchored to the front, with a small J hook bent in the unattached end, will make a secure latching mechanism (Fig. 7-7).

GETTING READY FOR SUMMER

You need to make arrangements for the comfort of rabbits in the hot months. Make sure the hutch walls, or the curtains on the building, can be easily opened during those warmish spells when the temperature may hit 80 degrees or above.

Refurbish cage wire that may have rusted during the damp winter weather when outdoor painting was out of the question. Non-toxic aluminum paint makes an excellent and inexpensive touch-up for those small areas that are just beginning to show signs of rust.

Fig. 7-7. A completed carrier.

Take the rabbits out of the cages needing repairs. Then, with a stiff wire brush, clean off the rusted area. This is a good time to make use of the propane torch, too. After burning off hair (and germs with it), brush off any ash residue. Paint those bad spots, or repair the whole cage, if desired. A new coat of paint makes the wire last that much longer.

While the outdoor hutch is empty, the exterior wood parts can also be repainted if needed. Paint the lower parts of legs on outdoor hutches with roofing tar or creosote to repel mites that originate in the soil.

Spring is also the best time to check the roofing to make sure there are no leaks to drip on rabbits and into nest boxes. Wet rabbits become sick rabbits.

Save any burlap bags, or some paper feed bags, to be soaked in cold water during very hot days. When the temperature soars over 80, soak some of the bags, fold them, and place them in the cages of due-to-kindle does. A wet sack to rest on can save the doe's life during heat waves.

As soon as the soil is warm enough for planting, put those insect repellent vine seeds in the ground on the west side of hutches if there isn't sufficient shade from trees. The choice of vines will depend on the area and what grows best.

Keep some extra water bowls handy and clean. A doe and a litter of eight can consume a gallon of water per day during hot weather. They will probably drink more during the night hours. If the water isn't there for them, they suffer from thirst. Meat production is slowed without ample water. This is one of the major reasons so many people have gone to automatic watering systems during the past few years.

Be sure that working bucks are in the coolest part of the rabbitry. Bucks go sterile from high temperatures more quickly than does. A wet sack to rest on doesn't harm the working buck.

Getting litters during the heat of summer can sometimes be a problem (Fig. 7-8). Breed as early in the morning as possible, or during the night hours. The hours between six and eight in the morning seem most productive, after the rabbits have enjoyed the cooler night hours.

Hot weather can slow down a good doe's milk production. Now is the time to give her a little boost with a good protein supplement or vitamins to help her create more milk for her young. Start a few days prior to kindling date, if protein supplement or vitamins aren't given on a regular basis, and continue through at least the first three

Fig. 7-8. Raising a litter during the heat of summer can cause problems. Preparing for those hot days ahead of time will make a doe's job a lot easier, thereby improving production.

weeks of nursing. After this, the young will be getting a part of their nourishment from pellets.

New babies in the nest can also suffer from heat. When the thermometer hits 85 or 90, the wise breeder places the young along with some of the nesting material and fur into a wire basket on top of the doe's cage. After the heat of day has passed, the young and the nesting materials are returned to the nest box where the doe can feed them.

When young, first litter does are due to kindle, stay close. Check on them frequently, as these young does become more excited than older, experienced does. They become overheated while in labor and the stress of kindling. It may be necessary to place them in a bottomless cage on the cool ground. Their place of confinement should be out of the way so they don't have to be moved again and are safe from animals like dogs and cats.

If one has a large enough freezer, plastic jugs filled with water may be frozen and placed in the cage with the rabbit during hot daytime hours. Just set the jug in the cage; the rabbit will know what to do with it.

Occasionally, there will be a rabbit that will suffer more from the heat, and more quickly, than you expected. If a rabbit is discovered with wetness showing around the mouth, dip it all the way to the neck into a pail of water (not very cold). Be sure that the fur becomes soaked all the way to the skin. Partial wetting can cause the rabbit to develop pneumonia. Place the wet rabbit back into its cage after wiping off excess water. Circulating air around it will cool the rabbit. No fans or drafts through cracks in a wall, though.

If a doe is seen with her front feet and legs resting in her water bowl, it's time to do something to relieve the heat. Give her the aforementioned wet sack, place her on the ground in a floorless cage, or give her a jug of frozen water to curl up around. Rabbits can and do produce good litters during hot weather. They don't tolerate heat too well, though, and will need help.

WINTER RABBIT RAISING

When you are raising rabbits for meat, it's important to keep them in production year-round. Here are some of the things we do for our winter rabbits.

Our rabbits are housed outside in hutches with three walls and the floor being made of wire, one solid wall, and a fiberglass roof. In winter we staple indoor-outdoor carpeting over the wire sides of the hutches for a windbreak. The lowest temperature here last year was 15 degrees below zero. Perhaps you couldn't get away with this kind of hutch if your temperatures are much colder. We've only lost one litter to the cold, and that was when we and the doe were both inexperienced.

We use all wood nest boxes 12 inches by 18 inches by 12 inches high with a 6 by 6 inch door set at the extreme top left of one 12 by 18 foot wall (Fig. 7-4). We make them so the top and bottom come off for easy cleaning. Cover the exposed edges of the door with aluminum flashing, stapled on, or else it will soon be chewed away. One week before the doe is due to kindle (or earlier if the doe starts carrying hay around in her mouth and looking anxious), we put the box in the cage and stuff it full of straw. Over the next few days, as the doe packs down the straw while she's making the nest, we add more straw. In really cold weather, the doe will make a tunnel in the straw that seems much too small for her to get into and the bunnies will be snug in the bottom. You won't even be able to see that there are bunnies; you'll have to reach in to check. Don't reach in too much with cold hands.

The doe will push any dead ones out toward the edge where you can see them. Put a brick or something similar on the cage floor in front of the nest box door about a week after kindling so the bunnies can get back into the nest should they fall out when they start getting more active. The height of the nest box door will keep them from falling out before they are hopping around enough to climb back in. We always check for bunnies out of the nest and return them when we are taking care of the rabbits twice a day.

At 15 degrees Fahrenheit it takes a lot of energy just to maintain a normal body temperature. It's hard for the doe to regain her strength enough to get bred if it's cold and she's been nursing a big litter. Therefore, we creep-feed Purina Bunny Chow beginning

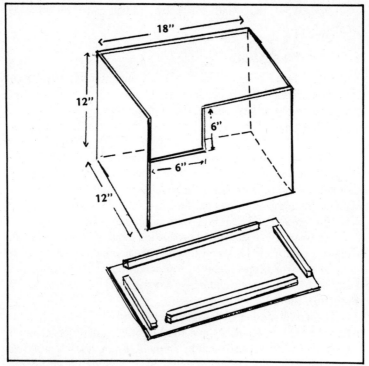

Fig. 7-9. The top of the nest box is made the same way as the bottom, with a gap in stop-strips at the door.

as soon as the bunnies are climbing out of the box, and we wean the bunnies at four weeks by removing them from the doe's cage. The doe will be producing a lot of milk at this time, so don't wean the whole litter at once. Take away all but two bunnies, then the next day or so take away one more, then the last one. We have divided a litter in half and left half with the doe for eight weeks and weaned the other half to Bunny Chow at four weeks. The bunnies that were weaned earlier were heavier at eight weeks when they were slaughtered. Bunnies also use a lot of energy in cold weather to keep warm, so be sure they have access to plenty of Bunny Chow until six weeks, then switch them to rabbit pellets or whatever you use for adult animals.

Be sure they get enough water. If you don't have automatic water, this means giving them water at least every 12 hours in a large enough volume to keep it liquid for several hours at a time. In the winter we don't rebreed our New Zealand White does until eight weeks after they kindle. This gives them time to get in good condition. When the weather is warmer, you can rebreed at six weeks.

BREEDING FOR STAMINA

To medicate or not to medicate is a question that often faces the rabbit breeder. There is a saying among long-time rabbit raisers that is some of the best advice the beginner can obtain, "The less medication needed in the rabbit herd, the better the herd gets along."

The novice breeder soon discovers that he has certain rabbits in the herd that seem never to need medications, regardless of weather, a rigorous breeding schedule, or whatever. The wise beginner recognizes this as hardiness—a stamina that permits the animal to enjoy good health in spite of outside factors.

You won't find a successful professional rabbit breeder wasting his herd's health and his own valuable time and income on trying to keep an ailment prone rabbit in top condition. This practice is a waste of time and money and is also a terrific hazard to the health of the rest of the herd.

The hardiest doe may be able to raise litters of lovely, healthy little rabbits, but she will need the help and protection of being housed next door to other healthy animals. Being housed next to a rabbit that catches a cold with every litter kindled or with every change in the weather subjects the hardy doe not only to the stress of her own kindling and nursing, but also exposes her to the cold germs sneezed or coughed into the air by the ailing doe. The rabbit with the cold will be isolated to prevent contamination of other rabbits, but the healthy doe may have already been exposed to the germs for several hours.

Unwise use of medications can destroy stamina in the hardiest, most disease-resistant herd. Use of antibiotics such as penicillin should be limited to those having the most severe infection and should be given for only short periods, never extending for weeks or months.

One effect is the immunization of the trouble-causing bacteria to the drug being administered. A more serious effect is the destruction of beneficial flora (bacteria) in the stomach and intestinal tract and even in the reproductive system.

Aside from the destruction of beneficial bacteria, some experienced breeders now feel that overuse of antibiotics may result in deafness in the animal treated. It is also known that certain drugs, used indiscriminately, can cause detrimental changes in the vaginal secretions of rabbit does, resulting in sterility and other breeding problems.

How do you go about keeping a herd healthy in spite of all the stresses of weather, breeding, and raising of young? Raisers with a great deal of experience say "breed in stamina." This is not as hard as it sounds, actually.

For instance, let's say you have purchased two does and one buck (a trio) from a reputable breeder. Chances are that of the two does, one will persistently have bunnies that stay healthier than the other doe's young. These hardy youngsters may or may not be prettier than the other doe's offspring. First, we work for a high rate of stamina, then we can go after the good looks.

Watch the litters of both does, sired by the same buck. When a bunny in either litter shows up with any sort of ailment, be sure to mark its dam's hutch card to that effect. When the fryers are ready for market or butchering for meat, look over both does' records. From the litter showing the fewest remarks about any sort of illnesses, choose two of the best built does and one buck.

Best build, production-wise, means a doe with a good deep rib cage, nice wide loin, and broad hips not pinched in at the tail. The broad, ample rump on the work doe makes it easier for her to kindle large litters of sound babies with fewer problems in delivering. The buck saved should be short-bodied, with good breadth and depth of body; have a fairly wide, short head; and should be active and energetic. Both does and bucks should have good thick fur; heavy pads on the feet; fine, straight ears; and clear, bright eyes.

When the young animals are mature, breed the buck back to his own dam, and the two young does back to their own sire. In this way, we find out just which side of the original pair contributed most of the stamina to the offspring. Once again, there will likely be one of the two young does that will do a better job of finishing out her litter than her sister.

Watch all three of these inbred litters for any signs of weaknesses or faults and ailments. If the litter from the mother-son breeding shows the most strength and stamina, then this is the litter from which new stock should be saved. If one of the father-daughter litters does better, then stock should be saved from that litter.

Inbreeding can be continued, with good rabbits, for three generations. Each successive generation should be watched very closely for weakness—floppy ears; weak legs that bow in or out; thin, skimpy fur on pads; narrow rib cages or hips; or any tendencies toward illness that might appear.

The third generation stock saved should be mated with unrelated or very distantly related rabbits, to prevent any drastic increase in possible genetic weaknesses. These fourth generation animals may then be mated back with the original pair, if desired, since they will be related to that first pair on only one side. After this fourth generation production, one should have enough stock to begin successful line-breeding to keep the good family traits originating with the first inbreedings.

Mating of aunts-nephews, uncles-nieces, and cousins (not of the first cousin generations) are line-breedings, and with careful planning, can be carried on indefinitely with success. Line-breeding is less risky than those inbreeding, but is slower. Also, if one starts out with this method, more beginning stock is required.

Now that you have a few fine, hardy rabbits that rarely if ever need any sort of medications, the trick is to keep them that way. No animal, and especially no heavy producing animal, can stay healthy without a balanced diet. The rabbit with a high rate of stamina seldom needs more than a nutritious diet, plenty of fresh clean water, and a little attention to sanitation to stay healthy.

A few things will need regular attention to prevent diseases. Ear canker caused by ear mites is not a disease if not neglected. Monthly applications of oil (vegetable, mineral or olive) in the ears and regular cage cleanings keep ear mites under control. A coccidiostat given in drinking water for three consecutive days a month, keeps liver coccidiosis from causing damage. A coccidiostat can be obtained from almost any feed store, pet shop, or drugstore having a veterinarian medicine department.

If the rabbit continues to have a cold or catches a fresh cold easily, it should be removed from the herd. No rabbit should ever be kept in the herd long enough for it to accumulate a matted, messy residue from its nose on the front legs and feet. This practice is deliberately asking for trouble with rabbits in adjoining cages and eventually for the entire herd.

If one of the rabbits you saved for breeding stock should suddenly show up with a slight cold, it should receive an intramuscular injection of a broad spectrum antibiotic such as Pen-Strep or Combiotic. One or possibly two shots should do the trick if the rabbit has had a balanced diet to keep up its stamina. Vitamins given in drinking water can help the doe through the stresses of kindling and nursing heavily if weather is a possible factor in her catching a cold.

Chapter 8

Bees

If you are a homesteader, but do not have colonies of bees on your premises, and no one in your family is allergic to bee stings, we urge you to give serious consideration to beginning such an enterprise. The bees will produce some honey, and this operation will give a challenging diversion from the routine day-to-day chores.

There are several ways to begin a beekeeping enterprise. You can buy a full size hive or nucleus that contains some combs of brood, and a laying queen, or you can start from the beginning by buying component parts. Put them together, then you introduce your bees and queen at the appropriate time and watch the colony grow. This latter technique is cheaper. It is also the best way to learn all the ramifications of beekeeping. Suppliers sell the beginners or hobbyist kit that contains all the necessary items to start. As the season progresses, you then buy honey supers and possibly an extractor.

You have another option. Suppliers also sell equipment to commercial beekeepers, and in some cases the quantity discount is substantial. If you pool your order with a friend or neighbor, you might be able to take advantage of the quantity discount.

To start your enterprise, you need a smoker, hive tool, veil, possibly white coveralls, and gloves. Start with a minimum of two colonies. While some might suggest no more than 10, try two or three the first year, then add more the second year. If you decide to end your adventure in beekeeping, you need not write it off as a total

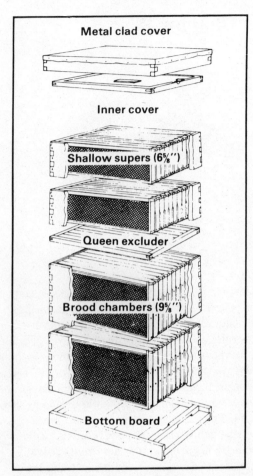

Fig. 8-1. A typical hive.

loss, as used beekeeping equipment does have value and can be sold.

The reason for starting two colonies is quite simple. If you have only one colony and the queen for some reason dies before she lays eggs, or the season is such that there is a cool or rainy period when the queen should mate, your colony is hopelessly queenless. If you have two colonies, the likelihood of this happening to both colonies is rather remote. If one colony becomes queenless, it can be united with the queen-right colony and you're still in business.

If you're not familiar with beekeeping equipment and terminology, Fig. 8-1 will help. Some beekeepers prefer to have deep brood chambers (9⅝ inches) and shallow supers (6⅝ inches) for honey. Each chamber or super holds 10 frames. You can use all shallow or

268

all deep equipment; however, a deep full super of honey in our opinion is too heavy to handle. If you decide to use all shallow equipment, you will need three shallows to take the place of two deep chambers. We personally prefer all shallow equipment because all parts (frames and chambers) are interchangeable. So for two colonies, you will need two top covers, two inner covers, two bottom boards, 15 shallow hive bodies (supers), 150 frames, and foundation.

You can build some of your own equipment. Be sure to make it to the standardized size and make it to exact dimensions. Some books have plans for standardized equipment. The United States Department of Agriculture and many state universities also have plans available. The other option is to buy or borrow one set and use it as a pattern.

It will cost you between $200 and $300 to get in the beekeeping business the first year and substantially less in the following years. Honey production is normally less the first year, not only because of mistakes you might make, but also bees will have to build new combs.

Average honey production varies from state to state, ranging from 50 to 125 pounds per colony. It's not unusual for a hobbyist to double or even triple his state average by intensive management. You can calculate your anticipated return by using your state average production and multiply it by the wholesale or retail price of honey, less container. Current wholesale price ranges between 40 to 50 cents per pound, retail 90¢ to $1.

PACKAGE BEES

There are several ways to obtain bees. First, you can obtain an established colony from a beekeeper. Second, you can hive a swarm. Third you can purchase a package of honey bees from a bee breeder in the south. For beginners, especially, we recommend the latter. The main advantage is that you will start with new equipment and young, gentle bees. More importantly, you can learn and grow with your package bees as they develop.

Package bees originate in the south because spring comes in time for colonies to build up so that they can be shipped to the northern states. They have three main purposes: strengthening weak colonies, pollination, and starting new colonies.

A package of bees simply consists of a wooden box with screen wire on two sides that is filled with worker bees, a queen in her own cage (somewhat resembling a match box), and a can of sugar syrup

(Fig. 8-2). Packages are sold by weight. The most popular weights are 2, 3, 4, and 5 pounds. Roughly, there are 3,500 bees to the pound. The 2-pound package would mostly be used for strengthening weak colonies. The 3-pound package would be used for establishing new colonies while the 4 and 5-pound packages would be used for purposes of pollination.

The key factor in determining what size package to purchase is the law of diminishing returns. You would not want too large a package for your purpose for the cost would be prohibitive. At the same time you would not want too few for the colony would not develop properly.

Hybrids and several races of bees are available. We strongly recommend that the Italian bee be purchased.

Order package bees as early as possible, about the time the first pollen and nectar are available. It takes about two months for a package colony to build up for the main honey flow.

Development of a package colony is as follows: three weeks will elapse from the time the queen begins to lay until young bees emerge from their cells. Two additional weeks are required for these young bees to reach the adult, honey-gathering age. As you can see, it takes at least five weeks for the package colony to really start to build up.

Before your package arrives, prepare for them. Have everything in readiness. Have your hive assembled, painted, and in place.

Fig. 8-2. A queen and her attendants in a shipping cage (photo by Jean Martin).

The frames should contain full sheets of foundation. At this time there should only be five frames in the hive. The entrance should be reduced to prevent robbing. You should also have a deep hive body available so that you can place a feeding jar above the colony after your bees have been installed. Lastly, you should have some sugar syrup available. This is made of 1 or 2 parts sugar to 1 part water and is placed in a jar with six or eight pin holes in the lid and inverted over the opening in the inner cover or frames. The extra hive body is put in place and the top is put on.

We discourage the use of an entrance feeder, also known as a boardman feeder. If there should be a cold snap, the bees would not be able to use an entrance feeder where they could use a feeder on the top. There is also the danger of robbing with the use of an entrance feeder. Most importantly, however, bees naturally go up, not down, to feed.

Upon arrival, place the bees in a cool, dark place such as a basement. At this time give them sugar syrup by painting it on the screen wire with a brush. Give them all they will take. This will make them easier to handle. If need be, the bees can be left in the package, without harm, for a day or two if they are fed.

To install bees, remove the lid by prying it off with your hive tool. Remove the queen cage and check to see if she is alive. Remove the metal or cardboard (if any) covering the candy end of the queen cage.

To assist the bees in releasing the queen, punch a small hole in the candy with a finishing nail. Hang the queen cage between the top bars in the middle of the frames. This is an indirect method of release.

If there is no candy but only a cork, the following procedure would apply. Remove the cork and suspend the cage as above. Close the hive as soon as possible after the bees are installed. This is a direct method of release.

The next step is to remove the feeder can. You can do this by tilting the cage so the can will slide out. Shake a few bees out on to the tops of the frames where the queen cage is hanging. This is done for two reasons. First, the bees will know where the queen is. Second, they will keep the queen warm and feed her.

After you have done this, you are ready to place the package itself, with the remaining bees, in the hive where five frames have been removed in order to accommodate them (Fig. 8-3). The bees will exit by themselves.

Another method of installing bees is to wet them down with

Fig. 8-3. The package bees are about to be introduced to their new hive (photo by Jean Martin).

sugar syrup, or water, and "pour" them onto the bottom of the hive. They will roll out like so many beans.

Two days later the hive should be opened and the queen cage checked to be sure that she has been released. If she has not been released, release her at this time and discard her cage. Also, discard the shipping cage if it was left in the hive at the time of installation.

Check to see if the bees have enough sugar syrup. Without doubt, it will have to be refilled at least every few days. Package bees that are not fed will starve to death.

Check the package colony again one to two weeks later. Examine them for eggs and capped brood cells.

Too often a beginning beekeeper wants to open the hive, look for the queen, and check the progress. Do not disturb the colony any more than described above for it will do more harm than good. At least it will disturb the bees and at worst the bees may "ball" the queen and kill her. This is the peculiar phenomenon where the bees completely surround her and attempt to pull her apart. The only thing that the beekeeper can do is close the hive and examine it a few days later to determine if the queen is still alive.

EXAMINING AND HANDLING BEES

As far as bee stings are concerned, there is no topical dressing or treatment that is effective against the swelling and pain. Many people recommend various treatments ranging from ammonia or tobacco juice to applications prepared by pharmacuetical companies. None of these will do any good. They are a waste of money, except for their placebo effect.

When you are stung, you are left with a puncture wound. As the skin is the first line of defense against infection, the puncture wound is sealed off immediately. Therefore, nothing can get in and nothing can get out. Furthermore, upon being stung, the bee venom enters the skin tissues within a matter of seconds, leaving little or no time to do anything.

Beginners often make the mistake of opening the hives too often. As a rule of thumb, each colony should receive a complete examination three or four times a year. In early spring, check for brood, pollen, and sufficient honey. During the main honey flow, check for signs of swarming and crowded conditions. The fall is the time to examine hives for wax moths, sufficient honey, and solid brood patterns. It's best to examine hives on a warm, sunny day when the bees have been active all morning. On cool, cloudy days bees tend to be irritable and cross.

Before examining the hive, put on the veil and light the smoker. The best smoker fuel is old burlap. Cut off a piece, light it, stuff it into the smoker, and squeeze the bellows. This will produce a good flame. Then stuff the remainder of the burlap into the smoker with the hive tool. You want cool smoke, not a flame.

Put on the gloves. You will soon learn to work without them. One should wear loose, light-colored clothing when working with bees. Avoid dark, fuzzy clothes made from materials such as wool and felt.

With hive tool in hand, blow a few puffs of smoke into the entrance of the hive (Fig. 8-4). Then move behind the hive or to one side, lift up the top, and blow in a few more puffs of smoke. Remove the top. If there is an inner cover, smoke the bees through the hole in the cover before removing it. The hives should be smoked enough to drive the bees down between the frames. If the bees' heads are sticking up in line with the tops of the frames, more smoke is needed. If, however, the bees become irritable after being smoked, close up the hive and wait for a better day.

The most important aspect of working with bees is handling the frames. Your best bet is to remove the first or second frame from the side of the hive first. The reason for this is that the queen is less likely to be in one of the side frames. She usually stays in the middle frames. Because the frames are close together, the queen stands a good chance of being killed or injured if she is in one of the first frames removed.

Lift the frames out by holding both ends of the top bar. Once the frame is removed from the hive, it can be held by the end of the left top bar and the corner of the right bottom bar. Holding the frame in this fashion allows you to pivot the frame while keeping it balanced. Take out one of the frames and lean it against the side of the hive. Use the hive tool to slide frames away from the one you want to remove. Taking one frame completely out of the hive will give you more working room.

When removing frames, try not to pinch or kill any of the bees. It is not the loss of a few bees that matters. What does create a problem is the fact that when a bee is crushed or attempts to sting, an alarm is given that the other bees sense, causing them to become cross. Also, when manipulating the frames, be sure to make deliberate and slow movements rather than quick ones. Other than

Fig. 8-4. Blow a few puffs of smoke into the hive (photo by Jean Martin).

performing a complete examination of the hive, it is not necessary to tear down a hive. You can simply pry up a super with the hive tool and take a peek.

When closing the hive, put the frames back in the order in which they were taken out, starting with the last frame first. Be sure there is an equal amount of space between each frame as you replace them. Reverse the frames as you replace them. The end of the frame that was facing the front of the hive should be facing the back when replaced. This helps to obtain a comb that is of uniform thickness. Bees tend to have thicker combs toward the front of the hive, while the comb in the same frame tends to be thinner toward the back.

It doesn't take long to become adept at handling bees and examining hives. Practice soon makes perfect, particularly if a set routine is always followed.

HELPING YOUR COLONY GROW

Judicious management of your colonies in the spring will pay off with honey later in the season. Production is influenced by colony size and climate. You cannot do much about the weather, but you can help your colony grow in size. By now natural pollen is available in fairly large quantities, and the queen is laying at a rate proportional to the size of the cluster.

Consider the priorities of bees within the colony. Rearing brood has the highest priority next to survival. Within a small colony the large majority of the workers will be relegated to rearing brood and other miscellaneous chores. On warm days some old workers will venture out in search of pollen and nectar. They won't go out if there is work to be done inside, so a small colony has a large percentage of its workers at home in comparison to a large colony. This is verified by actual field measurements and experiences of observant beekeepers. A strong colony always produces substantially larger quantities of honey than does a weak one.

To help build or continue building a strong colony, be sure you have plenty of space above the brood area. This can be done by timely manipulation of hive bodies (reversing chambers) or adding more brood boxes. If dandelion and early fruit bloom are yielding nectar and pollen and your colony is quite strong, watch for swarming. It is easy to detect the possibility of swarming by timely inspection. Check the swarm cells closely, at least every 10 days. If a swarm cell contains an egg or developing larvae, be sure to destroy it. Otherwise, you will have a swarm issue before your next inspection.

Some colonies appear to have a greater desire for or instinct to swarm than others. Overwintered colonies with old queens are more likely to swarm than are nuclei or package bees. While swarming is natural, it is not conducive to producing a large quantity of honey.

Producing queens in large numbers is a rather specialized business. As a hobbyist, you may want to try this as a new experience or raise enough for fall requeening. Raising queens is a rather simple operation provided you understand a few basic principles of honey bee biology and behavior.

The queen is a product of genetics and developmental environment. Some professional queen breeders will consider genetics by using controlled mating and/or instrument insemination, but the hobbyist can only select his best colony for queen stock.

A colony begins to raise a new queen when the old one is absent or in a weakened condition or begins swarm preparation. Swarming is the natural way in which a colony reproduces. Using swarms to increase your operation is expensive in terms of honey production, and the prudent beekeeper should try to manage colonies in a manner that prevents swarming. This is done by timing hive body manipulation or by adding additional brood chambers.

Whether a new queen is produced by swarming, supersedure, or any other means, her genetics is the same. Rather than destroying swarm cells, consider rearing some of them and producing a few extra queens. Whether you rescue a swarm cell or produce one under controlled conditions, the end result is identical.

All fertilized eggs have the potential of developing into a queen. If the larva is fed royal jelly for five days, she develops into a queen, and a worker develops from those larva fed royal jelly for two-and-one-half days. By experimentally manipulating the quantity of royal jelly given a larva during her lifetime, it is possible to produce a large, productive queen or small, inferior queen. Each queen has the maximum potential of developing up to 200 ovarioles (egg tubes). A queen that develops under conditions of inadequate or limited amount of royal jelly can develop as few as 125 egg tubes. This queen would probably be superseded earlier.

You may want to build one or several small mating hive bodies or "nuc boxes." These are simply tight fitting boxes that hold three or four standard frames—deep or shallow depending on the size of your equipment. One auger hole, ¾ to 1 inch, is enough. Some beekeepers use a standard hive body and divide it into three separate compartments, each with its own auger hole entrance, with one on each side and a third one on the end for the middle compartment. While one solid piece of plywood can be used as a bottom board, three separate covers are necessary so each compartment can be opened and inspected independently.

To obtain queens under controlled conditions, find and remove the queen from the hive from which you wish to raise some extra queens. You can take the queen with one frame—of sealed brood—one or two containing some honey and pollen plus attached bees, and place them in one of your nuc boxes. Workers won't abandon their queen. She will continue laying, but at a reduced rate.

Meanwhile back in the old hive, workers begin raising several new queens. To simplify your operation, you might note the location of eggs within this hive so when you recover the queen cells you need not re-examine each frame.

Keep in mind your new queen is scheduled to emerge in 14 days. Sometime after the new queen cells are sealed and before the first one emerges, remove the frame and attached bees and place it in a nuc box. Chances are you will have queen cells on one or several different frames, and each frame can be placed into a separate nuc box. You now can replace your old queen and the several frames of brood and bees back into the original colony. Be sure to check the entire colony and destroy any developing queen cells. If you miss one or several the colony may not accept their "old" queen. As a rule, if she is laying and has a good population of workers on the combs, they will accept her.

Save queen cells. You don't have to transfer an entire frame to a nuc box. With a sharp knife, remove the sealed queen cell and place it in a nuc box. A newly sealed queen cell is very delicate, but older cells are a little sturdier (Fig. 8-5). You could place a mature cell on its side in the nuc box, but it is better to suspend a newly sealed cell in its natural downward position. Damaged cells could produce a deformed queen, or she may even die before emerging.

If you know the age of your queen cell, you can predict her date of emergence. Within five or eight days she will take her mating flight. Most flights take place between 1:00 and 5:00 P.M. when the temperature is at least 68 degrees Fahrenheit or warmer. Two to four days after mating, she should begin laying eggs. If you note that all the larvae are workers as contrasted to drones, you now have a "tested" queen.

The weather following emergence is critical and actually be-

Fig. 8-5. Note the sealed queen cells in the center (photo by University of Wisconsin-Extension).

yond the control of the beekeeper. It is for this reason it is risky to try to raise queens in northern parts of the country in early spring. Beekeepers rely on queen producers in the south for early spring queens. If she cannot fly because of inclement weather to mate, she begins laying unfertilized eggs and is known as a drone layer.

FEEDING THE COLONY

The quickest way to feed a colony is to give it combs of honey from other colonies known to be disease-free. Colonies needing stores can also be fed on dry granulated beet or cane sugar, sugar syrup, or sugar candy. Bees can use dry granulated sugar if the humidity is relatively high.

Dry sugar can be fed in several different ways. The simplest way is to tip the hive backward sufficiently to pour the sugar through the entrance to the back of the bottom board. Another method is to put the sugar in a tray, or on an inner cover with the bee escape hole open, or on a piece of heavy paper over the top bars beneath the outer cover. The division board feeder can also be used to feed syrups or dry sugar when placed next to the cluster in the brood chamber. Syrups made of granulated sugar or diluted honey can be fed in friction-top pails or screw-cap jars inverted over the bee escape hole of an inner cover. Such containers are generally enclosed in an empty hive body with the outer cover on top.

One of the critical problems facing beekeepers, not to mention the bees, is the lack of pollen in early spring. Spring buildup is not possible without it. To remedy the situation, bees can be fed either a pollen substitute or a pollen supplement.

There is really no substitute for pollen (Fig. 8-6). We have made substitute mixtures, but found the bees did not take to it. We have found that the pollen supplements work very well and that the bees take to it very quickly. During July and August, we trap pollen with a couple of pollen traps located in one or two of the cornfields near the apiary. We put the trapped pollen in tin cans with tight fitting lids. The pollen is placed in the cans with alternate layers of granulated sugar, with a final layer of pollen on top. Sugar is hygroscopic in nature, removing moisture from the pollen, thereby preserving it. Pollen can be kept in this manner for two years.

INCREASING THE POPULATION OF COLONIES

Not all colonies increase in population equally fast, even with the best management. The differences may be due to a variety of causes. If some colonies have more stores than they need (thereby reducing the space available for brood rearing), combs of honey can be given to colonies needing more stores in exchange for empty combs. Similarly, if some hives contain more brood than average, they can be equalized by removing the combs of emerging brood,

Fig. 8-6. Bees like pollen (photo by John Caulk).

with most of the adhering workers, and giving them to weaker colonies, taking care not to transfer the queen.

The weakest colonies in the apiary should be assisted in this way only after all the others have been equalized; then they are given any frames of brood still available and are thus built up as rapidly as possible. Another way of equalizing is to shake bees from frames of brood in strong colonies in front of the entrances of those to be helped. The young bees go in and are accepted and the older bees return to their own hives. The queen must not be shaken in this way.

The advantages of having colonies develop at the same rate and reach the honey flow equally strong in bees are as follows. The colonies are ready for a given manipulation at the same time. Fewer hive bodies are needed than if strong colonies are given supers in accordance with their individual needs. When properly done, equalizing probably results in an actual increase in the total number of bees in the apiary, since every queen is more nearly capable of laying to her capacity and no queen is restricted by having only a small number of nurse bees to feed her and her brood. Less manipulation is necessary when the honey flow begins (especially in comb honey production) in sorting combs of brood and in reducing the brood to one hive body. The brood is compact, which is especially desirable in comb honey production.

The work of equalizing colonies is considerable, and the beekeeper must determine for himself if it is profitable. Inexperienced beekeepers tend to handle supers instead of combs when equalizing colony strength. Many beekeepers practice reversing the first and second stories, putting the first story above the second in order to provide more room for brood rearing. The second story is generally full of brood and honey at this early examination, while the lower brood chamber may have some brood, pollen, and some honey in the side combs. Reversing the position of the two chambers will cause the bees to transfer the honey to the combs above, and the movement of the honey will stimulate the bees to greater effort in rearing brood. The queen will go up into the second story more readily than she will go down.

Some colonies become much stronger than others in early spring, because of exceptional queens or drifting of bees between hives (Fig. 8-7). Many beekeepers equalize the strength of such colonies by exchanging the location of weak colonies with that of the stronger ones. The field force of the stronger colonies will augment that of the weaker colonies without dangerously reducing the

Fig. 8-7. Bees will take "cleansing flights" in early spring (photo by Jean Martin).

strength of the better colonies. The greater amount of emerging brood in the stronger colonies will soon replace the hive population. If they have an abundance of stores, brood rearing will continue without interruption. Since the field bees will be returning with pollen, nectar, or water, there will be no destructive fighting at the entrances of the interchanged colonies.

During the first thorough examination, all brood combs should be thoroughly inspected for brood diseases and proper steps taken to avoid the danger of spreading the diseases. If brood diseases are present in the apiary, the free movement of combs in equalization colonies is likely to transfer the disease from one colony to another. This should be avoided.

Stimulating Brood Rearing

In areas where there is a shortage of spring pollen and nectar plants or where there is need for the colony to build up more rapidly than under normal conditions, brood rearing can be stimulated by giving the colonies pollen cakes and by feeding with a dilute syrup. Feeding requires some manipulation of the colony that is not beneficial in inclement weather. Consequently, many beekeepers believe that by providing an abundance of food in the fall or by giving stores rapidly in the spring, the colony will receive all the necessary stimulus to brood rearing and that stimulative feeding is not desirable. Stimulative feeding may be unprofitable unless the extra bees produced can be used to advantage.

Stimulative feeding with medicated syrup is one way of eliminating American foulbrood or European foulbrood from lightly infected combs. It is also useful in preventing these diseases from getting started in the colonies. If bee diseases are a problem, it is good practice to have the colonies in the home yard or in nearby apiaries, so that they can receive proper attention with less travel expense. The same holds true for building up weak colonies. If stimulative feeding is practiced, it is usually best to feed the colonies in the evening. This prevents the bees from flying (as a result of the feeding), and therefore robbing does not get started. If there is a shortage of pollen, pollen cakes should be fed regularly until an adequate source of fresh pollen is available.

Spreading brood used to be practiced to a greater extent in past years, but it has many advantages under certain conditions, as well as disadvantages in others. If an empty comb is placed in the center of the brood area, the bees will attempt to cover all of the brood and, in so doing, will keep that part of the empty comb which intervenes warm enough so that the queen will lay therein. When this new brood is well started, the manipulation may be repeated and still more eggs will be laid.

While this is attractive in theory, great care should be used not to overextend the ability of the bees to keep all the combs and brood warm. If cold weather sets in, the cluster may contract and leave the outside combs of brood exposed to starvation. It is much safer to place the outside empty combs next to the outer combs of brood, permitting the queen to expand her egg laying as the internal conditions of the colony permit.

It is safe to say that stimulative feeding and the spreading of brood should not be practiced early in the spring. It should be confined to a period of not more than six weeks previous to the particular honey flow for which the beekeeper is building up his colonies. If the main crop is in midsummer, the beekeeper need not force his colonies in the spring unless he desires to make increases, shake package bees, or build up colony strength for early pollination services. If there are long intervals between honey flows, the beekeeper must see that brood rearing is at its best during the period of six to eight weeks before each flow.

Colonies expand rapidly in the spring when nectar and pollen sources are available. A majority of the colonies will require additional room, and frequently the application of warm control measures is necessary before the main honey flow starts. As a general rule, efforts should be made to prevent crowding in the brood nest by providing sufficient room for brood rearing and for the storage of early honey.

Making Room

In many parts of the country, strong colonies may make preparations to swarm during fruit bloom unless they are given more room. In California, this may occur during the last week in March or early in April. In eucalyptus areas along the coast, swarming preparations may be made in February. Consequently, the principles associated with colony expansion must be understood and additional room supplied as circumstances arise in the different parts of the country.

If the first and second stores were reversed early in the season, they can be reversed again as soon as the second story is filled with brood and honey and additional room is available below. If the colony population is crowded in the two hive bodies, a third hive body of drawn combs should be added above and without a queen excluder. If the weather is settled and there is no danger of chilling the brood, the third set of combs can be placed between the two brood chambers to break up any swarming preparations that may have been started. Top supering, however, is generally the rule when the third super is given. This permits the expansion of the brood nest.

If frames of foundation have to be given instead of drawn combs, place the super of foundation on top and raise two combs from the sides of the brood chamber to the center of the super. This will attract the bees to the super much quicker and cause them to draw out the foundation if incoming nectar is available in quantity. Once the combs are well started, young bees will be attracted from the chambers below and thus relieve crowding in the lower brood chamber.

Strong colonies may have brood in three or more stories, and each may be crowded with bees by late spring or early summer. Such colonies are in prime condition for the production of either comb or extracted honey and frequently go into the honey flow without making preparations to swarm, particularly if they are headed by young queens.

The conditions favorable to the rapid increase in the size of the colony in the spring may be restated as follows: a large number of young, vigorous workers, due to successful wintering and early brood rearing; a prolific queen; abundant stores properly located in the hive so as to be easily accessible to the bees; a prolific race or strain of bees; and good brood combs of worker cells in quantity sufficient for the needs of the colony.

CARING FOR COLONIES IN THE FALL

The care you give colonies in the fall can be crucial to your success the following year. Because of this, fall management is often considered the starting point of the beekeeper's year in order to provide strong, productive colonies to produce the next year's crop of honey.

Start preparations for wintering at the end of August or early part of September. For successful wintering, the colony must be headed by a good laying queen as a large number of young worker bees are needed to get through the winter. A failing queen should be replaced with a young, vigorous one. Only bees reared late in the fall are able to live until next spring.

Start your fall inspection while there is still some nectar coming in. This reduces the danger of robbing. After making sure the queen is strong, check for brood disease. If you suspect disease, contact your apiary inspector.

Many beekeepers winter their bees in a story-and-a-half or a two-story colony. Both hive arrangements are suitable if sufficient food is present. This depends on your geographic location. The colder the climate, the more food the bees require.

Provide an upper entrance. This is simply a 1-inch auger hole bored into the hive body above or below the hand hold—never in it (Fig. 8-8). Otherwise, when you go to lift the hive body, you will receive quite a surprise. An upper entrance not only facilitates moisture removal from the hive, but also prevents suffocation of bees if the lower entrance should become clogged with dead bees or snow during winter.

The ability of a colony of bees to regulate temperature allows them to withstand severe cold spells of −40 degrees Fahrenheit. Naturally, this is only possible if the bees act as a unit or a cluster. When the temperature falls below 50 degrees Fahrenheit or so, they form a cluster.

Clustering usually takes place in the lower part of the hive, and the bees move as a unit from comb to comb. The unit or cluster is connected with its various segments, located between the different combs, by a connective cluster. This connective cluster is very important because it provides for communication between the bees in the interspaces and facilitates ventilation.

The body temperature of the bees inside a winter cluster fluctuates between 68 degrees Fahrenheit and 97 degrees Fahrenheit regardless of the outside temperature. Bees generate heat through metabolic processes and control heat loss through contraction of the cluster (Fig. 8-9). The temperature of the bees making up the outer shell of the cluster is maintained at about 43 degrees Fahrenheit to 46 degrees Fahrenheit.

Brood rearing is usually discontinued late in the fall. When the temperature inside the cluster is increased to 92 degrees Fahrenheit to 96 degrees Fahrenheit early in January, brood rearing begins again. The amount of brood reared during this period depends not only on the ability of the cluster to regulate temperature, but also on the available pollen supply—the protein source of the bees.

Each colony should have enough honey and pollen to last until spring. A well-filled deep hive body with some empty space in the center combs provides enough stores for a strong colony wintered in two hive bodies. Colonies without sufficient honey should be given full combs or fed enough sugar syrup to make at least 40 pounds of stored food.

Bees winter best on combs that have been used for brood rearing. If possible, do not winter bees on all new honey combs. Be

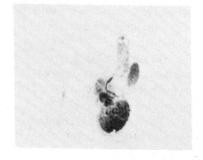

Fig. 8-8. An auger hole winter entrance.

Fig. 8-9. Bee-generated heat is melting snow off of the covers of these hives.

sure that any frames of foundation are replaced with drawn comb. Remove the excluder and all empty supers. If you have no other place to store empty combs, leave them on the hive above the inner cover with the center hole open. It is better to store combs where they cannot be damaged or blown over by the wind.

Weak or queenless colonies should be united with strong colonies having good queens. If you want to keep the individual small colonies rather than unite them, consider putting the small colony above a double division screen on a large colony. The heat will rise from the strong colony and keep the weak one warm.

As the weather becomes cooler at the end of summer, field mice look for warm places to spend the winter. A nest in the lower corner of a beehive is just such a place. For this reason it is necessary to use the ⅜-inch entrance reducer (Fig. 8-10). Do not make the entrance less than 4 inches wide or cover it with hardware cloth because the bees that die during the winter may block the entrance.

Cellar wintering of bees and wrapping of packing hives left outdoors were once common in many states. This is still practiced in parts of Canada. Most bees are now wintered without any special protection.

Wind protection is important to successful wintering. Shrubs, fences, or other artificial windbreaks help colonies survive by slowing heat loss. Snow may completely cover hives without damaging the bees, but the hives should not be located where water can collect. The winter apiary site should be on a slope or in an area where cold air flows away from the hives and will not collect around them. If your apiary location does not permit the sun to shine on the bees or is undesirable in other ways for wintering, plan to move the bees to a better location as soon as possible.

Losses of bees during winter are often high in spite of increasing knowledge about the biology and management of honey bees. Many bees of all ages die in the hive. Losses appear to be greater in very large and very small colonies compared with those of moderate size. It is not uncommon for more than half of the bees in a colony to die and for 10 percent or more of the colonies to die. Starvation, either from lack of honey or from the inability to reach the honey in extremely cold weather, is the most common cause of winter death of colonies.

Again, the amount of honey and pollen present in your colony during the winter will determine to a considerable extent the degree of successful wintering. Your colonies, depending on their strength, should enter the winter with about 40 to 80 pounds of honey. In addition, there should be several combs with pollen to allow for brood rearing activity the early part of the next year.

Never be concerned about leaving too much honey with your bees. Honey that is not consumed during the winter is not lost, but rather is utilized in the coming year. There are also considerable

Fig. 8-10. A mouse guard is shown in position.

differences between colonies in the amount of honey consumed during the winter. Generally, however, the amount of honey and pollen used is directly proportional to the strength of overwintered colonies.

SELECTIVE BREEDING

Livestock producers know the value of introducing new blood or new genes into their herd or flock. Bees are livestock and respond to selection and good breeding practices as do other animals. There are differences of opinion as to the value of and the best time to requeen.

There is no reason to requeen or buy outside queens to prevent inbreeding. While there is a possibility of it happening, the likelihood of it occurring in a specific colony or yard is extremely remote. When the queen leaves the colony to mate, she flies off and attracts drones from different colonies. These drones could come from yards five to 10 miles away. She also mates with anywhere from five to 10 drones. Nature has done well in protecting bees from inbreeding.

You can, however, improve a colony by requeening provided you obtain good queens. The queen is a product of genetics and her developmental environment. Queens and workers develop from fertilized eggs. All larvae are fed royal jelly for the first two-and-one-half days after hatching. Larvae destined to become queens continue to receive royal jelly for another two-and-one-half days while those who are to become workers are fed honey and some pollen.

The average female bee has the potential to develop about 200 egg tubules in the ovaries. Queens produced in a strong, vigorous colony develop about 180 tubules, while those which develop under less than ideal conditions develop about 125 tubules. Therefore, a queen with a good pedigree raised in a poor environment, for all practical purposes, is a questionable or poor queen. The converse could also be true. Some queen producers employ a system of grafting or double grafting. They simply transfer developing larvae to a cell containing a fresh supply of royal jelly—once or twice. This is a labor-intensive operation, but produces good queens. Once the cell is sealed, the queen's destiny is determined. It can be transferred to emergence cages, breeding nuclei, etc.

Performance testing in bees is not advanced to the degree it is in other livestock operations. Queen breeders have criteria on which they base their selection, but there are no standards. Therefore, when you purchase queens, you are relying on the reputation and integrity of the queen breeder. Some behavioral traits are inherited, especially temperament (some call it defensiveness). You don't have to keep bees very long to learn that there are rather striking differences in the temperament of bees. Over the years beekeepers in general and professional breeders tended to eliminate these, and now there are lines that are rather gentle. If you maintain colonies in suburbia, you will do well to requeen and keep requeening with queens from lines proven to be gentle. Honey production is not related to temperament, so there is no reason to tolerate uncooperative or highly temperamental bees.

Honey production is related to colony management, availability of nectar and pollen, and genetics of the queen. Requeening alone will not increase production, but will simplify one phase of management, namely swarm control. Swarming is influenced in part by conditions within the hive, photoperiod (length of day), and genetics. The instinct to swarm is also greater in older queens, and for that reason some beekeepers routinely requeen every year. Queens can live up to seven years, but most would probably be replaced in three years.

There are good arguments for fall requeening. They are cheaper; you can even produce your own almost anywhere in the United States. Producing spring queens in northern parts of the country is very risky. Those who requeen in the spring argue: why risk a new queen in a colony and lose both her and the colony during the winter?

An alternative is to requeen and divide in the spring. By feeding supplementary protein about eight weeks before natural pollen is available, it is possible to build a large, populous colony. Be sure to have queens available about mid-April or May, divide the colony, and introduce a marked queen (with a different color each year). You now have two colonies—one with a young queen and another with an old one. In the fall, eliminate your own queen and reunite the colonies into one single overwintering unit. This divided colony will be less interested in swarming than if it had remained as one unit.

Dividing a colony into two units is quite simple. Assume bees are in two deep hive bodies. All you need to do is rearrange combs of honey and pollen so each chamber has equal amounts. Place the old queen in the lower chamber, but this is not absolutely necessary. Cover this hive body with a piece of plywood, fiberboard, or inner cover with the hole tightly covered. Place the second hive body over this cover. You now have two colonies—one above the other. Be sure your top one has at least an auger hole opening. Your top colony is queenless; all you need to do is introduce your new young queen by the slow release technique. In three or four days she should be released and possibly laying. In several weeks you have two options. You can set the top chamber off on another bottom board and operate it as a separate colony, or leave it in place, but replace the separating cover with a queen excluder. You now have a two queen colony.

The two colonies will outproduce a single undivided overwintered colony, and the swarming instinct will be somewhat lessened but not necessarily eliminated. A two queen colony requires a somewhat different type of management; however, here, too, honey production also offsets costs.

At this time of the year you will notice considerable activity inside and on warm days outside the hive. If the weather cooperates, there usually is some nectar and pollen produced on early blooming plants, and bees are quite capable of taking care of their needs. Bees use about one cell of pollen and one of honey to raise one new bee. With brood rearing underway at a rapid rate, you will see little excess nectar and pollen stored because it is being used at a rapid rate. This is why a prudent beekeeper should keep an eye on the weather as well as the bees inside the colony. The reason is quite simple. Suppose you have a number of successive cold and/or rainy days when bees cannot fly. Should food reserves in the colony run low or out, you will have interrupted brood rearing. If this happens, your honey production will suffer later in the season.

Bees can continue to rear brood for about two weeks without pollen as they are able to draw on body reserves for protein. They don't have a reserve supply of energy (honey). If honey is exhausted, workers will throw out unsealed brood—and in time they could even starve. Therefore, be sure your colony has an adequate supply of honey. If in doubt, feed sugar or sugar syrup. While the standard supplementary feed is a heavy sugar syrup, granular sugar will also do in an emergency.

Some beekeepers keep a sack of sugar available for such emergencies, and colonies short on stores are fed immediately. Feeding dry sugar at this time is simple. Some just pour it over the bees, others put it on top of the inner cover, and others will place it on the bottom board. In the latter case you should block the lower entrance so it won't roll out the front. When the inside hive temperature is warm, bees will use the sugar and take care of their needs. If you help your colonies during periods of stress, they will return the favor by making excess honey for you later in the season.

HARVESTING AND PROCESSING THE CROP

Beekeepers often ask when and how often to remove honey from your colonies. This is perhaps the most frequently asked question from novice beekeepers. Never remove honey from the food or brood chamber. Bees must have sufficient stores to carry them through until the next honey flow, be it the fall honey flow or the spring honey flow. In the latter case, it is necessary to leave enough honey in the hive to carry the bees through the winter.

Basically, there are four types of honey: *section comb, chunk, extracted,* and *cut comb.* Section comb honey is produced by the bees in 1-pound wooden boxes. This type of honey production requires very strong colonies, a good nectar flow, and a great deal of swarm control management. We do not recommend this for beginners. Very few beekeepers produce this type of honey.

Chunk honey is very attractive and consists of chunks of honey placed in a jar with extracted (liquid) honey poured around it. It is easy to produce and sells well. Cut comb honey is similar, except that the chunks or pieces are put into plastic boxes.

Chunk honey is produced in shallow supers, and thin surplus foundation is used so it can be eaten. The comb is simply cut into pieces to fit the jar. This type of honey sells very well around here.

Extracted honey is simply honey that has been removed from the comb. As a rule, this is the only type of honey produced by commercial and hobbyist beekeepers. It is quick and easy to process, but its production requires extra equipment.

Comb honey should be removed as soon as it is completely capped over by the bees. If it is left on too long, the snow white cappings will become travel-stained, and you will not be able to sell it. The honey will be perfectly good, but the buying public—conditioned to dyed oranges and otherwise picture-perfect foods—will not buy it.

This is not the case with extracted honey, however, as the cappings are removed to extract the honey. The supers can be left on the entire spring, summer, and fall to be removed at your convenience. Extracted honey should not be removed until completely capped. If not capped, the liquid is still nectar and is often referred to as unripe or green honey. If green honey is removed from the colony, it will sour and ferment due to its high moisture content. If eaten, in severe cases it will cause diarrhea and cramps.

Fig. 8-11. A bee escape board.

The first step in harvesting the honey crop is removing the bees from the supers. If you have just a few colonies, simply shake the bees from the frames in front of the hive or brush them off with a soft bristle brush. If more than a few supers are involved, a bee escape should be used.

The bee escape is a metal device with two pairs of very small, sensitive springs, having points spaced so as to allow bees to exit, but not enter (Fig. 8-11). The bees can go in one direction, but not the other. The bee escape is placed in the hole in the inner cover beneath the super or supers of honey to be removed. Cover or plug all holes and cracks with rags or masking tape. If one small entrance is left, bees will come and rob the honey supers. If this occurs, you will be left with a nice set of empty combs. With the bee escape, it will take 24 to 48 hours for the supers to be cleared of bees. One disadvantage is that if you have outyards, this method necessitates two trips—one to put on the bee escapes and another to remove the honey.

We prefer not to place the bee escape in the hole of an inner cover. Rather we make a frame out of ¾-inch strips with a board down the middle in which we cut a hole for the bee escape. Next, we staple on screen wire on one side. This type of escape board is better than using the inner cover for several reasons. First, the screen wire allows communication between the bees, thus clearing them out more quickly. Second, there is better ventilation. During hot weather, the bees in the super stand a chance of suffocation. Third, if you wish, you can staple screen wire on both sides of the frame and then have a double screen that can not only be used for removing honey, but when the bee escape is covered, it can be used for queen rearing and queen introduction, too.

Another method of removing bees from supers is with a chemical such as Benzaldehyde. The chemical is applied to a *fume board*, which is nothing more than a frame covered with a cloth, placed on top of the supers that drives the bees down. Too much of the chemical will stupefy the bees and taint the honey. It can also be injurious to the beekeeper.

Large beekeepers use a piece of equipment called a bee blower. Although it is expensive, about $300, it is quick, safe, and very effective. The bee blower generates an air blast that blows the bees out.

Robbing is the stealing of honey from one colony by the bees of another colony. Bees protect what is theirs, and so a battle results and many bees are killed. Robbers can be recognized by their quick, bouncing movements about the corners of the hive or entrance. Robbing is caused by exposing honey anywhere in or near an apiary. This is particularly true when there is a dearth of nectar. If robbing should occur, close all hives, reduce the entrances, and remove all exposed honey.

Have all of your processing equipment ready before the honey flow commences. An extractor simply throws the honey out of the cells after they have been uncapped. A two frame extractor costs around $100. In addition, you'll need an electric uncapping knife, costing $25, and a cappings melter, costing $65. Also, you'll need strainers and a settling tank.

If you are a backlot beekeeper, you can process the honey in the kitchen, garage, or basement. Be sure that the area is bee-tight. If not, you will be driven away by bees trying to retrieve their honey. Above all, keep the equipment and working area clean. Cover the floor with several thicknesses of newspaper. As soon as honey drops on the floor, roll it up. If this is not done, honey will be tracked all over the floor in very short order.

Mount the extractor high enough, usually on a spare hive body, so that the honey can flow into a pail or bucket. Also, secure the extractor with turnbuckles so it will not wobble or walk around.

The uncapping can or melter should be located next to the extractor. Stack the supers next to it. Try to keep everything together to save steps. Furthermore, cut a piece of plywood the size of the bottom of a super and mount four ball bearings on the bottom of it. You can stack the supers on this and simply roll them around. Try anything to save your back.

Opposite the full supers should be a number of empty ones to receive the frames after they have been extracted. Be sure that these supers are sitting on some paper to catch dripping honey.

Now you are ready to uncap (Fig. 8-12). Take the first frame and rest the end-bar on the edge of the uncapping can or melter. Take the uncapping knife and start at the top with a sawing motion. The cappings will slice off in a solid sheet on shallow frames. Several passes will have to be made on the deep frames until you get the hang of it.

Next load the uncapped frames into the extractor. Try to equalize them by weight. This will help balance the extractor. Run the extractor until half of the honey is extracted from the first side. Reverse the combs so as to extract the other side. Reverse them again and extract the remaining honey on the first side.

Do not twirl the extractor too fast, or the combs may break. Be careful when straining the honey. You do not want to incorporate air bubbles into it. Small air bubbles will make the honey look cloudy. Honey should not be allowed to drip. Make a nylon bag and attach it to the gate of the extractor. It should be long enough to touch the bottom of the pail. By doing so, the honey will travel down the side of the bag into the pail, thus not incorporating air bubbles. As the pail is filled, you can raise the bag, but always keep the bottom of the bag submerged. As a further precaution, let the pail stand over night before bottling. This will allow the honey to settle.

If you do not wish to invest in an extractor, make a frame to fit your container and tack a piece of screen wire over it. Take a hot butcher knife and slice off the cappings. Invert the comb on the wire. Let it drip over night. Strain and now you have extracted honey.

In packaging section comb and chunk honey, be sure to use the whitest and best filled combs. Be sure to use crystal clear jars, preferably glass.

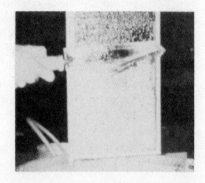

Fig. 8-12. Uncapping a deep frame of honey.

Extracted honey should be strained to remove bits of wax, bees and pollen. It should be allowed to settle overnight so that air bubbles and scum can rise to the top and be skimmed off. If this is not done, it will occur in the jars and give the honey a cloudy appearance, reducing the quality of your product.

The buying public looks for clarity, good flavor, good aroma, good body, and cleanliness. Also, all honey offered for sale must bear a label providing the buyer with the beekeeper's name, address, net weight, and bearing two very important words: "pure honey." This type of label more than likely will meet the regulations for interstate and intrastate sales.

BEESWAX

Wax is an important product or by-product of honey production. Under normal conditions, you can expect to produce about 1 to 2 pounds of wax per 100 pounds of extracted honey. (Wax and beeswax are often used interchangeably.)

Worker bees secrete wax with four glands located on the underside of the abdomen. She has to be about two weeks old before her glands are fully functional. Wax is only produced after her glands are fully functional and only when needed by the colony. Secreting wax is energy intensive; the colony will use about 8 pounds of honey to produce 1 pound of wax. Assuming honey wholsales for 50¢/pound, wax should bring $4, which is far from the price beekeepers receive today. Even at the current price of about $1.70/pound, it still is a valuable product. An effort should be made to salvage or recover as much as possible.

The highest grade of wax is obtained from the cappings. Don't mix capping wax with that which is recovered from rendering combs. There are a number of different ways to render wax. If you understand several basic points, you then can choose the method best suited to the size of your operation and available equipment.

Beeswax melts between 147 degrees and 149 degrees Fahrheit and has a density of 0.96. Density refers to the weight of a substance with reference to water. Beeswax is slightly lighter than water. Honey, with a density of 1.4, is heavier than water. Based on these facts, wax can be easily separated from honey by heat, and if you accurately control the temperature, the honey will not be damaged. The wax will float on either water or honey and, when solidified, it can be lifted off. There are many ways to heat the honey-wax mixture. If you use an open flame, be very careful—hot wax catches fire very easily. Extraneous materials in wax will cause it to foam when heated near boiling temperature of water, so don't leave wax on the stove unattended. A wax fire is difficult to extinguish.

There are many other materials in wax such as pollen, dust particles, propolis, etc. These materials (slum gum) usually settle to the bottom of the wax layer or bottom of the mold. If you sell your wax to a commercial processor, he will discount your wax depending on the amount of slum gum. You can also separate the slum gum from pure wax. Some processors will buy slum gum and pay you on the basis of recovered wax.

You may have a market for a small amount of clean wax to hobby shops or individuals who want it for some special projects. Beeswax can be easily washed. All you need to do is melt it in hot water, then pour it through a fine mesh strainer or cheese cloth when still hot, then allow it to solidify. You might have to repeat the process several time, but you will end up with clean beeswax.

While brood combs never wear out, there are occasions when you may want to recover the wax from old combs and put in a new foundation. Consider reworking brood comb with substantial drone comb (25 percent or more).

A shell of the cocoon is left in each cell after an adult bee emerges. The queen will lay another egg inside the "used" cell, so each cell may contain a number of cocoons. Although old brood comb is somewhat heavier than extracting comb, it won't contain much more wax, but enough to justify recovering it.

The number of combs and the available equipment will dictate the method you might use to render old combs.

1. Melt in hot water, then strain through wire screen or cheese cloth. You might repeat the process several times to recover a larger percentage of the wax.

2. Submerge below a wire screen in large kettle or tank of hot water. Upon melting, the wax floats to the surface and can be lifted off when cool.

3. Submerge in burlap sack in tank of hot water, then use same technique as in two.

4. Place in a heated wax press or centrifuge. This is generally used by large commercial operators.

5. Place combs in a solar extractor. A solar extractor is simply a glass covered box tilted in the direction of the sun, with a drain spout at the lower end so melted wax drains into a collution mold.

Regardless of the method you use to handle wax, keep capping wax separate from that recovered from old brood combs.

Index

A
African geese, 223
Amino acids, 68
Anconas, 207
Anemia, 76
Angora goats, 170
Antibiotics, 72
Antienterotoxemia, 149
Artificial insemination, 16, 90
Ashbrook, Paul, 185

B
Barbados Blackbelly sheep, 119
Bees, 267-296
Bees, examining, 272
Bees, feeding, 278
Bees, handling, 272
Bees, package, 269
Bees, selective breeding, 288
Bees, stimulating blood rearing, 282
Beeswax, 295
Bennett, Bob, 254
Blackface breeds of sheep, 110
Blindness, 77
Blue Andalusions, 208
Boars, 97
Boars, castrating the older, 99
Bowen, Godfrey, 166
Brahmas, 209
Breeding, 19
Breeding, after, 19
Breeding cows, 16
Breeding goats, 187
Breeding rabbits, 264
Brintanna Petite rabbits, 253
Buff geese, 223
Butchering pigs, 105

C
Caesarean, 22
Calories, 67
Calving, 18-22
Calving, abnormal, 21
Calving, before, 18
Calving, normal, 20
Calving, through, 18
Canadian geese, 223
Carnation Alber's Calf Manna, 33
Casting of wethers, 35
Castrating lambs, 153
Castrating pigs, 96

Cattle, 1-35
Cattle, Irish Dexter, 6
Cattle, Longhorn, 3
Cattle, Shorthorn, 4
Cellote, 236
Checkered Giant rabbits, 253
Cheviot sheep, 116
Chicken feeder, 238
Chinese geese, 223
Cochins, 209
Colonies, caring for in the fall, 284
Colonies, increasing the population, 279
Colonies of bees, 275
Colt training, 40
Combiotic, 93, 266
Comfrey, 73
Considine, Harvey, 185
Cornish, 210
Corriedale sheep, 113
Cows, 2
Cows, breeding, 16
Cows, milking by hand, 7
Cows, sources, 2
Cows, treating sick, 33
Crossbreds, 111
Crutch, 164

D
Deacons, 29
Dehorning goats, 199
Dehydration, 26
Diestrus, 91
Diet of geese, 223
Disbudding kids with a hot iron, 202
Docking lambs, 153
Doctoring calves, 25-28
Dorkings, 211
Dorset sheep, 110, 115
Draft horses, 36
Drenching goats, 181
Drenching sheep, 130
Drewry, Ken, 95
Dry period, 1
Dystocia, 21

E
Egyptian geese, 223
Emasculator, 99
Embden geese, 223
Energy and feeds, 56

Epiglottis, 181
Estrus, 91
Ewes, 109
Ewes, maintaining, 160
Ewes, managing pregnant, 140
Exercise yard for goats, 177

F
Farrier, 48
Farrowing house, 82
Feeder pigs, 79
Feeding lambs, 157
Feeding rack for sheep, 125
Fencing for sheep, 124
Finnsheep, 109, 121
Flemish Giant rabbits, 253
Fluorine poisoning, 78
Forest for hogs, 83
Foulbrood, 282

G
Galvayne's groove, 54
Gambrel, 101
Ganders, 225
Geese, 222
Geese, African, 223
Geese, Buff, 223
Geese, Canadian, 223
Geese, Chinese, 223
Geese, diet, 223
Geese, Egyptian, 223
Geese, Embden, 223
Geese, housing, 223
Geese, Pilgrim, 223
Geese, Sebastopol, 223
Geese, social life, 225
Geese, Toulouse, 223
Giant Chinchilla rabbits, 253
Gilts, 95
Goat barn, 176
Goats, 168-205
Goats, Angora, 170
Goats, breeding, 187
Goats, dehorning, 199
Goats, dehorning mature, 201
Goats, drenching, 181
Goats, exercise yard, 177
Goats, feeding program, 180
Goats, hay feeder, 176
Goats, Pygmy, 168

297

Goats, raising kid, 197
Goats, Saanen, 186
Goslings, 223

H
Hampshire sheep, 110, 113
Harnessing a horse, 43
Hatching eggs, 243
Hay feeder for goats, 176
Heifer, 1
Hogs, 61
Hogs, feed, 65
Hogs, feeder, 63
Hogs, forest, 83
Hogs, heat cycle, 91
Hogs, raising in summer, 82
Hogs, skinning, 101
Homasote, 236
Honey, 291
Honey, chunk, 291
Honey, cut comb, 291
Honey, extracted, 291
Honey, harvesting, 291
Honey, processing, 291
Honey, section comb, 291
Hoof care for horses, 48
Hoof rot, 50
Horses, 36-60
Horses, harnessing, 43
Horses, hoof care, 48
Horses, teeth, 51
Hot iron, 202
Housing for geese, 223
Huber, Maureen, 186
Huber, Samuel, 186
Hutches, 254
Hutches, wire, 254
Hutches, wood, 254
Hygrometer, 249
Hypoglycemia, 76

I
Incubator, 247
Infectious bovine rhinotrachaeitis (IBR), 25
Iodine deficiency, 78
Irish Dexter cattle, 6

J
Jersey Giants, 212

K
Kidding, 188
Kuhl, Gerry, 17

L
Lambing, 143
Lambs, 109
Lambs, castrating, 153
Lambs, docking, 153
Lambs, feeding, 157
Langshans, 213
Lice, 81
Loft requirements of pigeons, 232
Longhorns, 3
Loxon, 131
Lysol disinfectant, 99

M
Mastitis, 34
Mid-gestation, 18

Milk fever, 34
Milking machine, 12
Milking tips, 10
Minerals, 69
Minorcas, 214
Modern Games, 215
Mohair, 173
Muscovy ducks, 225

N
Netherland Dwarf rabbits, 253
New Hampshire Reds, 216
Niacin, 71
Nitrite poisoning, 78

O
Old English Games, 217
Orpingtons, 218
Osteomalacia, 76
Oxford sheep, 110

P
Package bees, 269
Pantothenic acid, 72
Parakeratosis, 77
Pen-Strep, 266
Phenothiazine, 131
Pigeons, 228
Pigeons, loft requirements, 232
Pigs, 61
Pigs, butchering, 105
Pigs, castrating, 96
Pigs, cold weather planning, 85
Pigs, feeder, 79
Pigs, plowing a garden, 87
Pilgrim geese, 223
Placenta, 196
Plymouth Rocks, 218
Post-lambing difficulties, 147
Poultry, 206-251
Proestrus, 91
Purina Bunny Chow, 262
Pygmy goats, 168

Q
Quonset hut, 235

R
Rabbits, 252-266
Rabbits, breeding, 264
Rabbits, breeding stock, 252
Rabbits, Brintanna Petites, 253
Rabbits, Checkered Giant, 253
Rabbits, Flemish Giant, 253
Rabbits, Giant Chincilla, 253
Rabbits, Netherland Dwarf, 253
Rabbits, raising in the summer, 259
Rabbits, raising in the winter, 262
Raising kid goats, 197
Ram, 134
Rambouillet sheep, 112
Rations for poultry, 238
Requeening, 289
Rhode Island Reds, 219
Riboflavin, 71
Rickets, 78
Roquefort cheese, 109

S
Saanen goats, 186
Salt deficiency, 77
Saw for dehorning, 200

Scalpel, 99
Scoop for dehorning, 200
Sebastopol geese, 223
Sebright, Sir John, 206
Selenium poisoning, 77
Seymour, John, 108
Seymour, Sally, 108
Shearing sheep, 162
Sheep, 109-167
Sheep, Barbados Blackbelly, 119
Sheep, blackface breeds, 110
Sheep, Cheviot, 116
Sheep, choosing the right breed, 112
Sheep, Corriedale, 113
Sheep, Dorset, 110, 115
Sheep, drenching, 130
Sheep, eating habits, 126
Sheep, feeding rack, 125
Sheep, fencing, 124
Sheep, Hampshire, 110, 113
Sheep, Oxford, 110
Sheep, Rambouillet, 112
Sheep, shearing, 162
Sheep, Shropshire, 116
Sheep, Southdown, 110, 117
Sheep, stopping foot rot, 133
Sheep, Suffolk, 110, 115
Sheep, teasing, 138
Sheep, trimming hooves, 132
Sheep, whiteface breeds, 110
Sheep, worming, 129
Shorthorns, 4
Shropshire sheep, 116
Skinning hogs, 101
Social life of geese, 225
Sources for cows, 2
South Dakota State University, 17
Southdown sheep, 110, 117
Sow, helping giving birth, 93
Sows, 62
Sphincter muscle, 11
Springer, 1
Squab, 228
Stomach tube, 25
Suffolk sheep, 110, 115
Supers, 292
Sussex, 220
Swarming, 289
Swine, 61-108

T
Teasing sheep, 138
Testosterone, 109
The Draft Horse Journal, 38
Thiabendazole, 131
Thrush, 50
Torsion of the uterus, 22
Total digestible nutrients (TDN), 56
Toulouse geese, 223
Tramisol, 131
Turkeys, 233

V
Vaseline petroleum jelly, 27
Vealers, 29
Vitamins, 71

W
Wallowing, 84
Whiteface breeds of sheep, 110
Worming sheep, 129
Wyandottes, 221